"十三五"国家重点出版物出版规划项目
卓越工程能力培养与工程教育专业认证系列规划教材
（电气工程及其自动化、自动化专业）

电气工程及其自动化专业英语

康忠健　张丽霞　郭　静　高春侠　编

机械工业出版社

本书从适应高等院校电气工程及其自动化专业英语的教学需求出发，兼顾内容的实用性和易用性，基本涵盖了电气工程及其自动化专业英语的相关知识。

本书共11章，内容分为三大部分。第一部分介绍由发电系统、输电系统、配电系统构成的电力系统一次系统，以及负责电力系统中测量、控制、保护的二次系统的相关内容；第二部分介绍一些目前比较热门的研究领域，包括运用在长距离、大容量输电情况下的、以电力电子技术为理论基础的高压直流输电技术和柔性交流输电技术，利用新能源发电的分布式发电系统，智能电网的组成及相关技术等；第三部分介绍电力系统及组成系统的电力设备的模拟仿真软件。

本书选材新颖，内容丰富，知识性强，注释详尽，语言精练，是高等院校电气工程及其自动化专业英语的适用教材，也可作为其他相近专业的参考用书，还可供有关技术人员选用。

图书在版编目（CIP）数据

电气工程及其自动化专业英语/康忠健等编. —北京：机械工业出版社，2021.9（2025.2重印）

"十三五"国家重点出版物出版规划项目 卓越工程能力培养与工程教育专业认证系列规划教材. 电气工程及其自动化、自动化专业）

ISBN 978-7-111-69586-8

Ⅰ.①电… Ⅱ.①康… Ⅲ.①电气工程-英语-高等学校-教材②自动化系统-英语-高等学校-教材 Ⅳ.①TM②TP2

中国版本图书馆CIP数据核字（2021）第231603号

机械工业出版社（北京市百万庄大街22号 邮政编码100037）
策划编辑：路乙达 责任编辑：路乙达 杨晓花
责任校对：王玉鑫 责任印制：邓 博
北京盛通数码印刷有限公司印刷
2025年2月第1版第3次印刷
184mm×260mm·18印张·445千字
标准书号：ISBN 978-7-111-69586-8
定价：59.00元

电话服务　　　　　　　　网络服务
客服电话：010-88361066　机　工　官　网：www.cmpbook.com
　　　　　010-88379833　机　工　官　博：weibo.com/cmp1952
　　　　　010-68326294　金　书　网：www.golden-book.com
封底无防伪标均为盗版　　机工教育服务网：www.cmpedu.com

"十三五"国家重点出版物出版规划项目
卓越工程能力培养与工程教育专业认证系列规划教材
（电气工程及其自动化、自动化专业）
编审委员会

主任委员

郑南宁　中国工程院 院士，西安交通大学 教授，中国工程教育专业认证协会电子信息与电气工程类专业认证分委员会 主任委员

副主任委员

汪槱生　中国工程院 院士，浙江大学 教授
胡敏强　东南大学 教授，教育部高等学校电气类专业教学指导委员会 主任委员
周东华　清华大学 教授，教育部高等学校自动化类专业教学指导委员会 主任委员
赵光宙　浙江大学 教授，中国机械工业教育协会自动化学科教学委员会 主任委员
章　兢　湖南大学 教授，中国工程教育专业认证协会电子信息与电气工程类专业认证分委员会 副主任委员
刘进军　西安交通大学 教授，教育部高等学校电气类专业教学指导委员会 副主任委员
戈宝军　哈尔滨理工大学 教授，教育部高等学校电气类专业教学指导委员会 副主任委员
吴晓蓓　南京理工大学 教授，教育部高等学校自动化类专业教学指导委员会 副主任委员
刘　丁　西安理工大学 教授，教育部高等学校自动化类专业教学指导委员会 副主任委员
廖瑞金　重庆大学 教授，教育部高等学校电气类专业教学指导委员会 副主任委员
尹项根　华中科技大学 教授，教育部高等学校电气类专业教学指导委员会 副主任委员
李少远　上海交通大学 教授，教育部高等学校自动化类专业教学指导委员会 副主任委员
林　松　机械工业出版社 编审 副社长

委员（按姓氏笔画排序）

于海生	青岛大学 教授	王　平	重庆邮电大学 教授
王　超	天津大学 教授	王再英	西安科技大学 教授
王志华	中国电工技术学会 教授级高级工程师	王明彦	哈尔滨工业大学 教授
		王保家	机械工业出版社 编审
王美玲	北京理工大学 教授	韦　钢	上海电力大学 教授
艾　欣	华北电力大学 教授	李　炜	兰州理工大学 教授
吴在军	东南大学 教授	吴成东	东北大学 教授
吴美平	国防科技大学 教授	谷　宇	北京科技大学 教授
汪贵平	长安大学 教授	宋建成	太原理工大学 教授
张　涛	清华大学 教授	张卫平	北方工业大学 教授
张恒旭	山东大学 教授	张晓华	大连理工大学 教授
黄云志	合肥工业大学 教授	蔡述庭	广东工业大学 教授
穆　钢	东北电力大学 教授	鞠　平	河海大学 教授

序

工程教育在我国高等教育中占有重要地位，高素质工程科技人才是支撑产业转型升级、实施国家重大发展战略的重要保障。当前，世界范围内新一轮科技革命和产业变革加速进行，以新技术、新业态、新产业、新模式为特点的新经济蓬勃发展，迫切需要培养、造就一大批多样化、创新型卓越工程科技人才。目前，我国高等工程教育规模世界第一。我国工科本科在校生约占我国本科在校生总数的1/3，近年来我国每年工科本科毕业生约占世界总数的1/3以上。如何保证和提高高等工程教育质量，如何适应国家战略需求和企业需要，一直受到教育界、工程界和社会各方面的关注。多年以来，我国一直致力于提高高等教育的质量，组织并实施了多项重大工程，包括卓越工程师教育培养计划（以下简称卓越计划）、工程教育专业认证和新工科建设等。

卓越计划的主要任务是探索建立高校与行业企业联合培养人才的新机制，创新工程教育人才培养模式，建设高水平工程教育教师队伍，扩大工程教育的对外开放。计划实施以来，各相关部门建立了协同育人机制。卓越计划要求试点专业要大力改革课程体系和教学形式，依据卓越计划培养标准，遵循工程的集成与创新特征，以强化工程实践能力、工程设计能力与工程创新能力为核心，重构课程体系和教学内容；加强跨专业、跨学科的复合型人才培养；着力推动基于问题的学习、基于项目的学习、基于案例的学习等多种研究性学习方法，加强学生创新能力训练，"真刀真枪"做毕业设计。卓越计划实施以来，培养了一批获得行业认可、具备很好的国际视野和创新能力、适应经济社会发展需要的各类型高质量人才，教育培养模式改革创新取得突破，教师队伍建设初见成效，为卓越计划的后续实施和最终目标的达成奠定了坚实基础。各高校以卓越计划为突破口，逐渐形成各具特色的人才培养模式。

2016年6月2日，我国正式成为工程教育"华盛顿协议"第18个成员，标志着我国工程教育真正融入世界工程教育，人才培养质量开始与其他成员达到了实质等效，同时，也为以后我国参加国际工程师认证奠定了基础，为我国工程师走向世界创造了条件。专业认证把以学生为中心、以产出为导向和持续改进作为三大基本理念，与传统的内容驱动、重视投入的教育形成了鲜明对比，是一种教育范式的革新。通过专业认证，把先进的教育理念引入了我国工程教育，有力地推动了我国工程教育专业教学改革，逐步引导我国高等工程教育实现从课程导向向产出导向转变、从以教师为中心向以学生为中心转变、从质量监控向持续改进转变。

在实施卓越计划和开展工程教育专业认证的过程中，许多高校的电气工程及其自动化、自动化专业结合自身的办学特色，引入先进的教育理念，在专业建设、人才培养模式、教学内容、教学方法、课程建设等方面积极开展教学改革，取得了较好的效果，建设了一大批优质课程。为了将这些优秀的教学改革经验和教学内容推广给广大高校，中国工程教育专业认证协会电子信息与电气工程类专业认证分委员会、教育部高等学校电气类专业教学指导委员会、教育部高等学校自动化类专业教学指导委员会、中国机械工业教育协会自动化学科教学委员会、中国机械工业教育协会电气工程及其自动化学科教学委员会联合组织规划了"卓越工程能力培养与工程教育专业认证系列规划教材（电气工程及其自动化、自动化专业）"。本套教材通过国家新闻出版广电总局的评审，入选了"十三五"国家重点图书。本套教材密切联系行业和市场需求，以学生工程能力培养为主线，以教育培养优秀工程师为目标，突出学生工程理念、工程思维和工程能力的培养。本套教材在广泛吸纳相关学校在"卓越工程师教育培养计划"实施和工程教育专业认证过程中的经验和成果的基础上，针对目前同类教材存在的内容滞后、与工程脱节等问题，紧密结合工程应用和行业企业需求，突出实际工程案例，强化学生工程能力的教育培养，积极进行教材内容、结构、体系和展现形式的改革。

经过全体教材编审委员会委员和编者的努力，本套教材陆续跟读者见面了。由于时间紧迫，各校相关专业教学改革推进的程度不同，本套教材还存在许多问题。希望各位老师对本套教材多提宝贵意见，以使教材内容不断完善提高。也希望通过本套教材在高校的推广使用，促进我国高等工程教育教学质量的提高，为实现高等教育的内涵式发展贡献一份力量。

卓越工程能力培养与工程教育专业认证系列规划教材
（电气工程及其自动化、自动化专业）
编审委员会

前　言

工业作为国民经济的支柱产业对经济的发展起着至关重要的作用。随着科技的进步和市场经济的日新月异，工业飞速发展，电气工程及其自动化技术在现代工业中的地位和作用也越来越突出。电气工程及其自动化技术作为一门具有开发潜力的综合性学科，它与多种学科的高新技术相互衔接，在现代工业生产的各相关领域得到普遍应用。同时，随着国际交流与合作，以及大量先进的电气工程及其自动化技术设备引进的日益增多，电气工程及其自动化专业英语的应用也越来越广泛。学习专业英语是了解国外学术及行业最新动态、掌握专业领域最新知识的基础，是高等院校电气工程及其自动化专业学生和相关领域工程技术人员必备的专业技能。

为了满足高等院校电气工程及其自动化专业英语的教学需求，特此编写了《电气工程及其自动化专业英语》。本书选材于国内外最新出版的教科书、专著、外文期刊等，基本涵盖了电气工程及其自动化专业英语的相关知识。全书共分为11章，包括电力系统中的发电系统、输电系统、配电系统，以及电力系统中的测量和控制、高压直流输电技术、柔性交流输电系统、智能电网、广域测量保护系统、电力系统仿真等电气工程及其自动化领域专业知识。通过对本书的学习，可以了解电气专业技术中外用语表述的相互联系，提升电气专业英语的阅读和应用能力，为今后通过阅读外文文献、著作，获取电气专业领域的前沿知识、开展国际交流与合作打下基础。

本书第1~4章、第11章由康忠健编写，第5~8章由张丽霞编写，第9章由郭静编写，第10章由高春侠编写，全书由张丽霞统稿。

编者在本书编写过程中参考了大量相关技术资料，吸取了许多专家和同仁的宝贵经验，在此深表谢意。

由于时间仓促、水平有限，错误之处在所难免，恳请广大读者和同行批评指正。

编　者

目 录
CONTENTS

序
前言

Chapter 1　Power Generation System 1
　1.1　Introduction 1
　1.2　Generating Electric Currents 2
　1.3　Electric Power Station 4
　　1.3.1　Steam Generator 4
　　1.3.2　Other Turbine Generators 6
　1.4　Synchronous Generator Voltage Control 7
　　1.4.1　Excitation Systems 7
　　1.4.2　Types of Excitation Systems 9
　　1.4.3　Modelling of Excitation System Components 10
　1.5　Motor Applications in Power Plants 11
　　1.5.1　Categories of Electrical Motors 12
　　1.5.2　Speed Control of DC Motors 18
　　1.5.3　Speed Control of AC Motors 26

Chapter 2　Transmission System 32
　2.1　Introduction 32
　2.2　Transmission Lines 32
　　2.2.1　Overhead Lines 32
　　2.2.2　Underground Cables 33
　2.3　Transformers 35
　　2.3.1　Introduction 35
　　2.3.2　Representation of Two-Winding Transformers 36
　　2.3.3　Representation of Three-Winding Transformers 39
　　2.3.4　Phase-Shifting Transformers 40
　2.4　Extra High Voltage (EHV) Transmission 41
　　2.4.1　Introduction 41
　　2.4.2　EHV Transmission in China 42
　　2.4.3　Future of EHV Transmission Grid in China 44
　　2.4.4　EHV Grid Effects on the Electricity Markets 46

Chapter 3　Distribution Network 49

3.1　Introduction ……………………………………………………………… 49
3.2　Distribution Systems Structure …………………………………………… 50
　　3.2.1　Primary Distribution Systems …………………………………… 50
　　3.2.2　Consumer Distribution Systems ………………………………… 51
　　3.2.3　Commercial and Industrial Installations ………………………… 52
　　3.2.4　Consumer Loop System ………………………………………… 54
　　3.2.5　Secondary High-Voltage Distribution …………………………… 55
　　3.2.6　Secondary Ties Loop System …………………………………… 55
3.3　Load in Distribution System ……………………………………………… 57
　　3.3.1　Static Load Models ……………………………………………… 57
　　3.3.2　Dynamic Load Models …………………………………………… 59
3.4　Distribution Network Automation System ……………………………… 62
　　3.4.1　Introduction of the Distribution Network Automation System … 62
　　3.4.2　Control Type Selection for Distribution Network Automation … 63
　　3.4.3　Switch Type and Control Method in Distribution Network …… 64
　　3.4.4　Operation of Tie Switch in Distribution Network ……………… 65
　　3.4.5　Components Selections in DAS ………………………………… 66
　　3.4.6　Simplified Supervising System in DAS ………………………… 67

Chapter 4　Measurement and Control in Power System ………………… 70

4.1　Introduction ……………………………………………………………… 70
4.2　SCADA System in Power System ……………………………………… 73
　　4.2.1　SCADA Overview ……………………………………………… 73
　　4.2.2　SCADA Architectures …………………………………………… 78
4.3　Power System Control …………………………………………………… 82
　　4.3.1　Key Concepts in Protection and Control ……………………… 82
　　4.3.2　Power Generation Control ……………………………………… 84
　　4.3.3　Power Substations Control ……………………………………… 85
　　4.3.4　Electric Utility Control Center ………………………………… 87
　　4.3.5　The Future of The Electric Utility Control Center …………… 89

Chapter 5　High Voltage Direct Current(HVDC) Transmission ………… 107

5.1　Introduction ……………………………………………………………… 107
5.2　Converters in HVDC …………………………………………………… 110
　　5.2.1　Solid-State Power Device in HVDC …………………………… 110
　　5.2.2　Rectifying Systems in HVDC Systems ………………………… 111
　　5.2.3　Inverter in HVDC ……………………………………………… 117
5.3　Application of HVDC …………………………………………………… 122
　　5.3.1　Corona Discharge ……………………………………………… 123

	5.3.2	AC Network Interconnections	124
	5.3.3	Renewable Electricity Superhighways	124
	5.3.4	Voltage Sourced Converters (VSC)	125
5.4	HVDC in China		125

Chapter 6　Flexible AC Transmission (FACT) Systems　129

- 6.1　Introduction　129
- 6.2　Static Var Compensator (SVC)　141
  - 6.2.1　Introduction　141
  - 6.2.2　Principles of Var Compensation　142
  - 6.2.3　AC Controller-Based Structures　143
  - 6.2.4　Active Compensator Topologies　145
- 6.3　Uniform Power Flow Controller (UPFC)　145
  - 6.3.1　Introduction　145
  - 6.3.2　Basic Structure of the Unified Power Flow Controller　147
- 6.4　Bridge-Type Fault Current Controller (FCC)　149
  - 6.4.1　Introduction　149
  - 6.4.2　Bridge-Type Fault Current Controller with a Bias Power Supply　149

Chapter 7　Renewable Energy and Distributed Generation　154

- 7.1　Introduction　154
- 7.2　Wind Power　155
  - 7.2.1　Wind Turbine Basics　156
  - 7.2.2　A Walk Around the Wind Turbine Control Loops　159
  - 7.2.3　Modeling and Control of Wind Farms　162
- 7.3　Solar Power　164
  - 7.3.1　Solar Cells　164
  - 7.3.2　Integrated Solar Home System　166
  - 7.3.3　PV Utilization in China　169
  - 7.3.4　Wind-Solar Hybrid System　173
- 7.4　Nuclear Power and Other　174
  - 7.4.1　How Nuclear Power Works　174
  - 7.4.2　Nuclear Fission: The Heart of the Reactor　175
  - 7.4.3　Pros and Cons of Nuclear Power　176

Chapter 8　Smart Power Grid　180

- 8.1　Introduction　180
- 8.2　Smart Substation　181
  - 8.2.1　Introduction　181
  - 8.2.2　Primary Equipment Design　182

8.2.3	Secondary Equipment Design	188
8.2.4	Benefits of the New Substation Design	194
8.2.5	Future Substation Design	194

8.3 The Smart Grid 203
    8.3.1 Introduction 203
    8.3.2 Smart Grid Definitions 205
    8.3.3 Smart Grid Components 205
    8.3.4 Smart Grid in China 209

Chapter 9 Wide Area Measurement Protection System 214

9.1 Introduction 214
9.2 Phasor Measurement Unit (PMU) and Its Application 214
    9.2.1 Defination of Phasor 214
    9.2.2 Sources of Synchronization 216
    9.2.3 Phasor Measuring Units 217
    9.2.4 PMU Potential Applications 218
9.3 Wide-Area Protection System 222
    9.3.1 Introduction 222
    9.3.2 Requirements on Protection Compared to SCADA/EMS 224
    9.3.3 Architectures of Wide-Area Protection System 224
    9.3.4 Discussion 227

Chapter 10 Electronic Devices Simulation 230

10.1 Introduction 230
10.2 Simulation Program with Integrated Circuit Emphasis (PSPICE) 231
    10.2.1 Introduction of PSPICE Fundamentals 231
    10.2.2 Element Statements 231
    10.2.3 Design Manager 232
    10.2.4 Device Model 236
    10.2.5 Library File 240
10.3 MATLAB/Simulink 240
    10.3.1 Introduction of MATLAB 240
    10.3.2 SimPowerSystems Block Libraries 241

Chapter 11 Power System Simulation Softwares 250

11.1 Introduction 250
11.2 Offline Simulation Softwares 250
    11.2.1 Power System Simulation for Engineering (PSS/E) 250
    11.2.2 PowerFactory (DIgSILENT) 251
    11.2.3 Power Systems CAD (PSCAD) 260

11.2.4	Power System Analysis Toolbox (PSAT)	262
11.3	Real-Time Digital Simulator (RTDS)	269
11.3.1	Introduction	269
11.3.2	RTDS Hardware	269
11.3.3	RTDS Software	272

Reference 276

Chapter 1
Power Generation System

1.1 Introduction

Electricity generation is the process of generating electric energy from other forms of energy. The fundamental principles of electricity generation were discovered during the 1820s and early 1830s by the British scientist Michael Faraday. His basic method is still used today: Electricity is generated by the movement of a loop of wire, or disc of copper between the poles of a magnet.

For electric utilities, it is the first process in the delivery of electricity to consumers. The other processes, electricity transmission, distribution, and electrical power storage and recovery using pumped storage methods are normally carried out by the electric power industry.

Electricity is most often generated at a power station by electromechanical generators, primarily driven by heat engines fueled by chemical combustion or nuclear fission, but also by other means such as the kinetic energy of flowing water and wind. There are many other technologies that can be and are used to generate electricity such as solar photovoltaics and geothermal power.

Centralized power generation became possible when it was recognized that alternating current power lines can transport electricity at very low costs across great distances by taking advantage of the ability to raise and lower the voltage using power transformers.

Electricity has been generated at central stations since 1881. The first power plant were run on water power or coal, and today we rely mainly on coal, nuclear, natural gas, hydroelectric, and petroleum with a small amount from solar energy, tidal harnesses, wind generators, and geothermal sources.

There are seven fundamental methods of directly transforming other forms of energy into electrical energy:

1) Static electricity, from the physical separation and transport of charge (e. g. triboelectric effect and lightning).

2) Electromagnetic induction, where an electrical generator, dynamo or alternator transforms kinetic energy (energy of motion) into electricity.

3) Electrochemistry, the direct transformation of chemical energy into electricity, as in a battery, fuel cell or nerve impulse.

4) Photoelectric effect, the transformation of light into electrical energy, as in solar cells.

5) Thermoelectric effect, direct conversion of temperature differences to electricity, as in

thermocouples, thermopiles, and thermionic converters.

6) Piezoelectric effect, from the mechanical strain of electrically anisotropic molecules or crystals.

7) Nuclear transformation, the creation and acceleration of charged particles (e. g. betavoltaics or alpha particle emission).

Static electricity was the first form discovered and investigated, and the electrostatic generator is still used even in modern devices such as the Van de Graaff generator and MHD generators. Electrons are mechanically separated and transported to increase their electric potential.

Almost all commercial electrical generation is done using electromagnetic induction, in which mechanical energy forces an electrical generator to rotate. There are many different methods of developing the mechanical energy, including heat engines, hydro, wind and tidal power.

The direct conversion of nuclear potential energy to electricity by beta decay is used only on a small scale. In a full-size nuclear power plant, the heat of a nuclear reaction is used to run a heat engine. This drives a generator, which converts mechanical energy into electricity by magnetic induction.

Most electric generation is driven by heat engines. The combustion of fossil fuels supplies most of the heat to these engines, with a significant fraction from nuclear fission and some from renewable sources. The modern steam turbine (invented by Sir Charles Parsons in 1884) currently generates about 80% of the electric power in the world using a variety of heat sources.

1.2 Generating Electric Currents

In this section, we will see that a change in the magnetic flux passing through a transformer's secondary coil causes current to flow in that coil. Since the magnetic flux through the secondary coil changes whenever the current through the primary coil changes, an alternating current in the transformer's primary coil induces an alternating current in its secondary coil.

But there's another way to change the magnetic flux passing through a coil of wire: move the magnetic flux. That's how a generator works. Whenever a magnet moves past a coil of wire or a coil of wire moves past a magnet, the flux through the coil changes and current flows in the coil and its circuit.

Most generators use rotary motion to produce electricity. The generator shown in Figs. 1.1a and 1.2 has a permanent magnet that spins between two fixed coils of wire. As the magnet spins, its magnetic flux lines sweep through the two coils and drive a current through them. This current experiences a voltage rise as it passes through the coils and a voltage drop as it passes through the light bulb, so it transfers power from the generator to the light bulb.

The iron core inside each coil extends the magnet's flux lines so that they are sure to sweep through the coil each time a pole of the magnet passes by. These cores are temporarily magnetized by the nearby magnet and effectively increase its length (Fig. 1.1a). Without the iron cores, most of the rotating magnet's flux lines would bend around before passing through the entire coil (Fig. 1.1b) and the generator would be less effective in producing electricity.

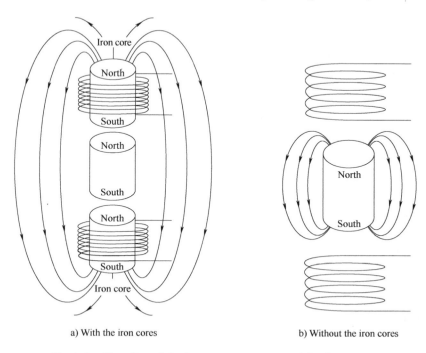

a) With the iron cores b) Without the iron cores

Fig. 1.1 The effect of the iron core on pass way of flux lines cores

A generator of this type produces an alternating current in the circuit it powers. This current flows in one direction as the magnet's north pole approaches a coil and in the opposite direction as the south pole approaches it. To generate the 60 Hz alternating current used in the United States, the generator must turn 60 times each second so that the current completes one full cycle of reversals every 1/60th of a second. In Europe, the generator must turn 50 times each second to supply 50 Hz alternating current. The generators throughout the continent-wide power distribution networks all turn together in perfect synchronization. That way, power can be redirected within each network so that any generator can provide the power consumed by any user.

Some devices require direct current electric power. The automobile is a good example. It generates DC electric power to charge its battery and to run its headlights, ignition system, and other electric components. While this power is actually produced by an AC generator or alternator, the car uses special electronic switches to send current from the alternator one way through its electric system. While the current in the alternator's coils reverses, the current through the car's electric system always travels in one direction.

Because large permanent magnets are extremely expensive, most industrial generators actually use iron-core electromagnets instead. These rotating electromagnets drive currents through generator coils just as effectively as permanent magnets would. Although these electromagnets consume some electric power, they are much more cost effective than real permanent magnets.

Since the electric current extracts energy from the generator, something must do work on it to keep it turning. We can see why this work is needed by looking at Fig. 1.2. As the permanent magnet turns its north pole toward the iron core above it, that core temporarily becomes magnetic with its south pole down. Opposite poles are then near one another and the permanent magnet is attracted toward the iron core. This attraction does (positive) work on the turning permanent magnet as the two poles approach but it also does negative work on the permanent magnet as the two poles separate. Overall, the iron core does zero net work on the permanent magnet.

But when the generator's coils are connected as part of a complete circuit, the currents induced in those coils make them magnetic, too. As required by Lenz's law, each magnetized coil repels the approaching permanent magnet. This effect is identical to the one we observed in electrodynamically levitated trains.

An approaching magnet is always repelled by the currents it induces. While a magnetized coil repels the turning permanent magnet, both as it approaches and as it leaves, the net work done by this repulsion isn't necessarily zero. If the current passing through the coils

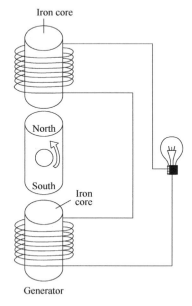

Fig. 1.2 A generator works by sweeping magnetic flux through wire coils

diminishes during the time between the permanent magnet's approach and its departure, the coil will repel the permanent magnet more strongly as it approaches than as it leaves and the net work done on the permanent magnet will be negative. Some of the permanent magnet's energy will be transferred to the electric current, which will deliver it to the devices in the circuit. The more power these devices extract from the circuit, the more work the permanent magnet must do to keep the current flowing.

1.3 Electric Power Station

1.3.1 Steam Generator

Since the permanent magnet loses energy to the electric current, something must exert a torque on it to keep it turning. While small generators are often turned by hand cranks, pedals, or internal combustion engines, most large power plant generators are driven by steam turbines. But instead of deriving their power from burning aviation fuel, power plant turbines operate on steam (Fig. 1.3).

A turbine resembles a fan run backward. While a fan uses rotary motion to push air from low pressure to high pressure, a turbine uses the flow of steam from high pressure to low pressure to propel a rotary motion. High-pressure steam exerts unbalanced pressures on the turbine blades and those blades rotate away from the steam. Since the steam's force and the direction of motion are both in approximately the same direction, the steam does work on the turbine and provides the ordered energy needed to spin the generator's permanent magnet.

This ordered energy comes from the steam's thermal energy. Thermal energy can't be converted directly into ordered energy. The steam turbine must and does operate as a heat engine, converting a

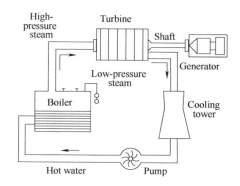

Fig. 1.3 Steam generator working diagram

limited amount of thermal energy into ordered energy as heat flows from a hotter object to a colder object. In the steam turbine, the hotter object is the high-pressure steam and the colder object is the outside air or water. As the steam's heat flows toward the colder world around, nature and thermodynamics allow us to extract some of that heat as ordered mechanical energy. A steam turbine is ideally suited to electric power generation because both involve rotary motions. The shaft of the turbine, on which its blades are fastened, experiences a torque from the steam and transfers that torque to the generator.

The turbine and the generator's permanent magnet rotate together, with power flowing from the high-pressure steam, to the turbine blades, to the shaft, to the permanent magnet, and finally to the electric current flowing through the generator (Fig. 1.4). The turbine spins at a steady, regulated rate to produce 60 Hz alternating current in the United States and 50 Hz in Europe and China.

Fig. 1.4 A high-pressure steam turbine drives this electric generator, producing roughly 66 MW of electric power

Steam is produced in a boiler, where thermal energy is added to liquid water until its molecules separate into a dense, high-pressure gas. The hotter the steam, the higher its pressure and the more effective it is at spinning the turbine. This thermal energy can come from burned fossil fuels, from nuclear reactors, or from sunlight.

As the steam flows through the turbine and does work on the rotating blades, its pressure and temperature drop. By the time the steam leaves the turbine, it has cooled considerably and its pressure is only slightly above atmospheric. It's time to return it to the boiler for reuse. But the steam must first be converted back into water because the work required to pump steam into the boiler is proportional to its volume. Turning the steam into dense liquid water reduces that work enormously.

The low-pressure steam flows through a cooling tower, where it gives up heat to the surrounding air. Often that heat is used to evaporate additional water, which then condenses in the cooler air above the tower as a plume of white mist. Once the steam has given up enough heat, it condenses into water and can be returned to the boiler. Many power plants are built near large bodies of water, which also receive some of the steam's waste heat. And a few modern power plants use this waste heat for other industrial or commercial purposes such as heating buildings.

Most electric power plants are thermal power plants. They have a number of components in common and are an interesting study in the various forms and changes of energy necessary to produce electricity.

(1) Boiler Unit

Almost all of power plants operate by heating water in a boiler unit into super heated steam at very high pressures. The source of heat from combustion reactions may vary in fossil fuel plants from the source of fuels such as coal, oil, or natural gas. Biomass or waste plant parts may also be used as a source of fuels. In some areas solid waste incinerators are also used as a source of heat. All of these sources of fuels result in varying amounts of air pollution, as well as, the carbon dioxide (a gas implicated in global warming problems). In a nuclear power plant, the fission chain reaction of splitting nuclei provides the source of heat.

(2) Turbine-Generator

The super heated steam is used to spin the blades of a turbine, which in turn is used in the generator to turn a coil of wires within a circular arrangement of magnets. The rotating coil of wire in the magnets results in the generation of electricity.

(3) Cooling Water

After the steam travels through the turbine, it must be cooled and condensed back into liquid water to start the cycle over again. Cooling water can be obtained from a nearby river or lake. The water is returned to the body of water 10-20 degrees higher in temperature than the intake water. Alternate method is to use a very tall cooling tower, where the evaporation of water falling through the tower provides the cooling effect.

1.3.2 Other Turbine Generators

All turbines are driven by fluid as an intermediate energy carrier. Many of the heat engines just mentioned are turbines. Other types of turbines can be driven by wind or falling water. Sources include:

(1) Steam

Water is boiled by nuclear fission, the burning of fossil fuels (coal, natural gas, or petroleum). In hot gas (gas turbine), turbines are driven directly by gases produced by the combustion of natural gas or oil. Combined cycle gas turbine plants are driven by both steam and natural gas. They generate power by burning natural gas in a gas turbine and use residual heat to generate additional electricity from steam. These plants offer efficiencies of up to 60%.

The steam generated by

1) Biomass.

2) The sun as the heat source: solar parabolic troughs and solar power towers concentrate sunlight to heat a heat transfer fluid, which is then used to produce steam.

3) Geothermal power. Either steam under pressure emerges from the ground and drives a turbine or hot water evaporates a low boiling liquid to create vapour to drive a turbine.

4) Ocean thermal energy conversion (OTEC): uses the small difference between cooler deep and warmer surface ocean waters to run a heat engine, usually a turbine.

(2) Other Renewable Sources

1) Water (hydroelectric): Turbine blades are acted upon by flowing water, produced by hydroelectric dams or tidal forces.

2) Wind: Most wind turbines generate electricity from naturally occurring wind. Solar updraft towers use wind that is artificially produced inside the chimney by heating it with sunlight, and are more properly seen as forms of solar thermal energy.

Small electricity generators are often powered by reciprocating engines burning diesel, biogas or natural gas. Diesel engines are often used for back up generation, usually at low voltages. However most large power grids also use diesel generators, originally provided as emergency back up for a specific facility such as a hospital, to feed power into the grid during certain circumstances. Biogas is often combusted where it is produced, such as a landfill or wastewater treatment plant, with a reciprocating engine or a microturbine, which is a small gas turbine.

1.4 Synchronous Generator Voltage Control

1.4.1 Excitation Systems

The basic function of an excitation system is to provide direct current to the synchronous machine field winding. In addition, the excitation system performs control and protective functions which are essential to the satisfactory performance of the power system by controlling the field voltage and thereby the field current.

The control functions include the control of voltage and reactive power flow, and the enhancement of system stability. The protective functions ensure that the capability limits of the synchronous machine, excitation system, and other equipment are not exceeded.

This chapter describes the characteristics and modelling of different types of synchronous generator excitation systems. In addition, it discusses dynamic performance criteria and provides definitions of related terms which are useful in the identification and specification of excitation system requirements. These serve as useful references to utilities, manufacturers, and system analysts by establishing a common nomenclature, by standardizing models, and by providing guides for specifications and testing. Models and terminologies used in this chapter largely conform to these publications.

The performance requirements of the excitation system are determined by considerations of the synchronous generator as well as the power system.

The basic requirement is that the excitation system supplies and automatically adjusts the field current of the synchronous generator to maintain the terminal voltage as the output power varies within the continuous capability of the generator. Margins for temperature variations, component failures, emergency overrating, etc., must be factored in when the steady-state power rating is determined. Normally, the exciter rating varies from 2.0 to 3.5kW/MV · A for generator rating.

In addition, the excitation system must be able to respond to transient disturbances with field forcing consistent with the generator instantaneous and short-term capabilities. The generator capabilities in this regard are limited by several factors: rotor insulation failure due to high field voltage, rotor heating due to high field current, stator heating due to high armature current loading, core and heating during underexcited operation, and heating due to excess flux (V/Hz). The thermal limits have time-dependent characteristics, and the short-term overload capability of the generators may extend from 15 to 60 seconds. To ensure the best utilization of the excitation system, it should be capable of meeting the system needs by taking full advantage of the generator's short-term capabilities without exceeding their limits.

From the power system viewpoint, the excitation system should contribute to effective control of voltage and enhancement of system stability. It should be capable of responding rapidly to a disturbance so as to enhance transient stability, and of modulating the generator field so as to enhance small-signal stability.

Historically, the role of the excitation system in enhancing power system performance has been growing continually. Early excitation systems were controlled manually to maintain the desired generator terminal voltage and reactive power loading. When the voltage control was first automated, it was very slow, basically filling the role of an alert operator. In the early 1920s, the potential for enhancing small-signal and transient stability through use of continuous and fast-acting regulators was recognized. Greater interest in the design of excitation systems developed, and exciters and voltage regulators with faster response were soon introduced to the industry. Excitation systems have since undergone continuous evolution. In the early 1960s, the role of the excitation system was expanded by using auxiliary stabilizing signals, in addition to the terminal voltage error signal, to control the field voltage to damp system oscillations. This part of excitation control is referred to as the power system stabilizer. Modern excitation systems are capable of providing practically instantaneous response with high ceiling voltages. The combination of high field-forcing capability and the use of auxiliary stabilizing signals contribute to substantial enhancement of the overall system dynamic performance.

To fulfill the above roles satisfactorily, the excitation system must satisfy the following requirements:

1) Meet specified response criteria.

2) Provide limiting and protective functions as required to prevent damage to itself, the generator, and other equipment.

3) Meet specified requirements for operating flexibility.

4) Meet the desired reliability arid availability, by incorporating the necessary level of redundancy and internal fault detection and isolation capability.

1.4.2 Types of Excitation Systems

Excitation systems have taken many forms over the years of their evolution. They may be classified into the following two broad categories based on the excitation power source used:

1) DC excitation systems.

2) AC excitation systems.

1. DC Excitation Systems

The excitation systems of this category utilize DC generators as sources of excitation power and provide current to the rotor of the synchronous machine through slip rings. The exciter may be driven by a motor or the shaft of the generator. It may be either self-excited or separately excited. When separately excited, the exciter field is supplied by a pilot exciter comprising a permanent magnet generator.

DC excitation systems represent early systems, spanning the years from the 1920s to the 1960s. They lost favour in the mid-1960s and were superseded by AC exciters.

The voltage regulators for such systems range all the way from the early non-continuously acting rheostatic type to the later systems utilizing many stages of magnetic amplifiers and rotating amplifiers.

DC excitation systems are gradually disappearing, as many older systems are being replaced by AC or static type systems. In some cases, the voltage regulators alone have been replaced by modern solid-state electronic regulators. As many of the DC excitation systems are still in service, they still require modelling in stability studies.

Fig. 1.5 shows a simplified schematic representation of a typical DC excitation system with an amplidyne voltage regulator. It consists of a DC commutator exciter which supplies direct current to the main generator field through slip rings. The exciter field is controlled by an amplidyne.

2. AC Excitation Systems

The excitation systems of this category utilize alternators (AC machines) as sources of the main generator excitation power. Usually, the exciter is on the same shaft as the turbine generator. The AC output of the exciter is rectified by either controlled or non-controlled rectifiers to produce the direct current needed for the generator field. The rectifiers may be stationary or rotating.

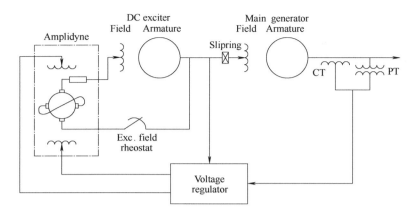

Fig. 1.5　DC excitation system with amplidyne voltage regulator

The early AC excitation systems used a combination of magnetic and rotating amplifiers as regulators. Most new systems use electronic amplifier regulators.

Thus AC excitation systems can take many forms depending on the rectifier arrangement, method of exciter output control, and source of excitation for the exciter. The following is a description of different forms of AC excitation systems in use.

1.4.3　Modelling of Excitation System Components

The basic elements which form different types of excitation systems are the DC exciters (self or separately excited); AC exciters; rectifiers (controlled or non-controlled); magnetic rotating, or electronic amplifiers; excitation system stabilizing feedback circuits; signal sensing and processing circuits. We describe here models for these individual elements. In the next section we will consider modelling of complete excitation systems.

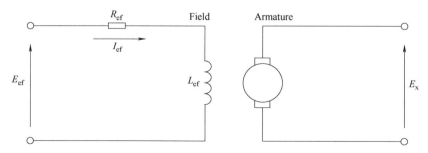

Fig. 1.6　Separately excited DC exciter

The circuit model of the exciter is shown in Fig. 1.6. For the exciter field circuit, we write

$$E_{ef} = R_{ef}I_{ef} + \frac{d\Psi}{dt}$$

$$\Psi = L_{ef}I_{ef}$$

Neglecting field leakage, the exciter output voltage E_x is given by

$$E_x = K_x \Psi$$

where K_x depends on the speed and winding configuration of the exciter armature.

The output voltage E_x is a nonlinear function of the exciter field current I_{ef} due to magnetic saturation. The voltage E_x is also affected by the load on the exciter. The common practice in DC exciter modelling is to account for saturation and load regulation approximately by combining the two effects and using the constant-resistance load-saturation curve, as shown in Fig. 1.7.

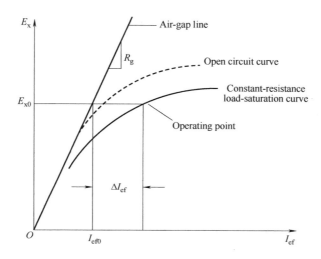

Fig. 1.7 Exciter load-saturation curve

The air-gap line is tangent to the lower linear portion of the open circuit saturation curve. Let R_g be the slope of the air-gap line and ΔI_{ef} denote the departure of the load saturation curve from the air-gap line. From Fig. 1.7, we write

$$I_{ef} = \frac{E_x}{R_g} + \Delta I_{ef}$$

The above equation gives the relationship between the output E_x and the input voltage E_{ef}. A convenient per unit system for this equation is one with base values of E_x and I_{ef} chosen to be equal to those values required to give rated synchronous machine voltage on the air-gap line.

1.5 Motor Applications in Power Plants

The use of electrical motors has increased for home appliances and industrial and commercial applications for driving machines and sophisticated equipment. Many machines used in electric power plants and automated industrial equipment now require precise control. Thus motor design and complexity has changed since early DC motors which were used primarily with railroad trains. Motor control methods have now become more critical to the efficient and effective operation of machines and equipment. Such innovations as servo-control systems and industrial robots have led to new developments in motor design.

Our complex system of transportation has also had an impact on the use of electrical machines. Automobiles and other means of ground transportation use electrical motors for starting and generators for their battery charging systems. There has recently been emphasis in the development of electric motor-driven automobiles. Aircraft use electrical machines in ways similar to automobiles. However, they also use sophisticated synchro and servo-controlled machines while in operation.

1.5.1 Categories of Electrical Motors

1. Types of DC Motors

Electromechanical energy-conversion devices that are characterized by DC are more complicated than the AC type. In addition to a field winding and armature winding, a third component is needed to serve the function of converting the induced AC armature voltage into DC voltage. Basically the device is a mechanical rectifier and is called a commutator.

Appearing in Fig. 1.8 are the principal features of the DC machine. The stator consists of an unlaminated ferromagnetic material equipped with a protruding structure around which coils are wrapped. The flow of direct current through the coils establishes a magnetic field distribution along the periphery of the air gap in much the same manner as occurs in the rotor of the synchronous machine. Hence in the DC machine the field winding is located on the stator. It follows then that the armature winding is on the rotor. The rotor is composed of a laminated core, which is slotted to accommodate the armature winding. It also contains the commutator—a series of copper segments insulted from one another and arranged in cylindrical fashion. Riding on the commutator are appropriately placed carbon rushes while serve to conduct direct current to or from the armature winding depending upon whether motor or generator action is taking place.

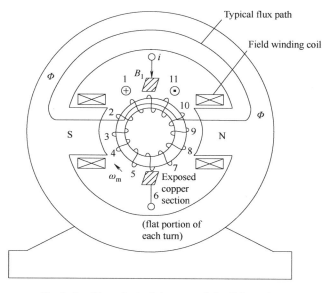

Fig. 1.8 The principal features of the DC machine

The types of commercially available DC motors basically fall into four categories: permanent-magnet DC motors, series-wound DC motors, shunt-wound DC motors, and compound-wound DC motors. Each of these motors has different characteristics due to its basic circuit arrangement and physical properties.

(1) Permanent-Magnet DC Motors

The permanent-magnet DC motor, shown in Fig. 1.9, is constructed in the same manner as its DC generator counterpart. The permanent-magnet motor is used for low-torque applications. When this type of motor is set, the DC power supply is connected directly to the armature conductors through the brush/commutator assembly. The magnetic field is produced by permanent magnets mounted on the stator. The rotor of permanent magnet motors is a wound armature.

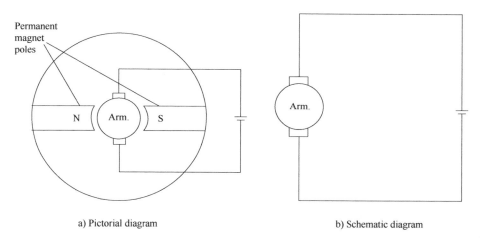

a) Pictorial diagram　　　　　　　　　　　　b) Schematic diagram

Fig. 1.9　Permanent-magnet DC motor

This type of motor ordinarily uses either alnico or ceramic permanent DC motor rather than field coils. The alnico magnets are used with high horsepower applications. Ceramic magnets are ordinarily used for low horsepower slow-speed motors. Ceramic are highly resistant to demagnetization, yet they are relatively low in magnetic-flux level. The magnets are usually mounted in the motor fame and then magnetized prior to the insertion of the armature.

(2) Series-Wound DC Motors

The manner in which the armature and field circuits of a DC motors are connected determines its basic characteristics. Each of the types of DC motors is similar in construction to the type of DC generator that corresponds to it. The only difference, in most cases, is that the generator acts as a voltage source while the motor functions as a mechanical power conversion device.

The series-wound motor, shown in Fig. 1.10, has the armature and field circuits connected in a series arrangement. There is only one path for current to flow from the DC voltage source. Therefore, the field is wound of relatively few turns of large diameter wire, giving the field a low resistance. Changes in load applied to the motor shaft cause changes in the current through the field. If the mechanical load increases, the current also increases. The increased current creates a

stronger magnetic field. The speed of a series motor varies from very fast at no load to very slow at heavy loads. Since large currents may flow through the low-resistance field, the series motor produces a high-torque output. Series motors are used where heavy loads must be moved and speed regulation is not important. A typical application is for automobile starter motors.

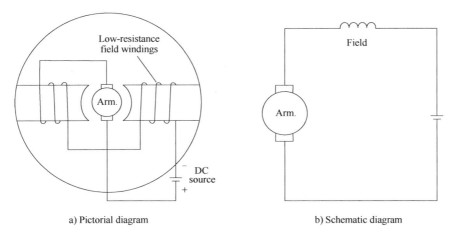

Fig. 1.10 Series-wound DC motor

(3) Shunt-Wound DC Motors

Shunt-wound DC motors are more commonly used than any other types of DC motor (Fig. 1.11). The field current may be varied by placing a variable resistance in series with the field windings. Since the current in the field circuit is low, a low-wattage rheostat may be used to vary the speed of the motor due to the variation in field resistance. As the field resistance increases, the field current will decrease. A decrease in field current reduces the strength of the electromagnetic field. When the field flux is decreased, the armature will rotate faster, due to reduced magnetic-field interaction. Thus the speed of a DC shunt motor may be easily varied by using a field rheostat.

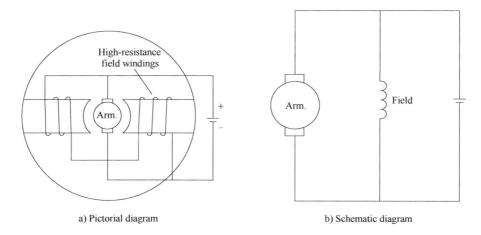

Fig. 1.11 Shunt-wound DC motor

The shunt-wound DC motor has very good speed regulation. The speed does decrease slightly when the load increases due to the increase in voltage drop across the armature. Due to its good speed regulation characteristic and its ease of speed control, the DC shunt motor is commonly used for industrial applications. Many types of variable-speed machine tools are driven by DC shunt motors.

(4) **Compound-Wound DC Motors**

The compound-wound DC motors shown in Fig. 1.12 have two sets of field windings, one in series with the armature and one in parallel. This motor combines the desirable characteristics of the series-and shunt-wound motors. There are two methods of connecting compound motors: cumulative and differential. A cumulative compound DC motor has series and shunt fields that aid each other. A differential compound DC motors have series and shunt fields that oppose each other. There are also two ways in which the series windings are placed in the circuit. One method is called a short shunt, in which the shunt field is placed across the armature. The long-shunt method has the shunt field winding placed across both the armature and the series field.

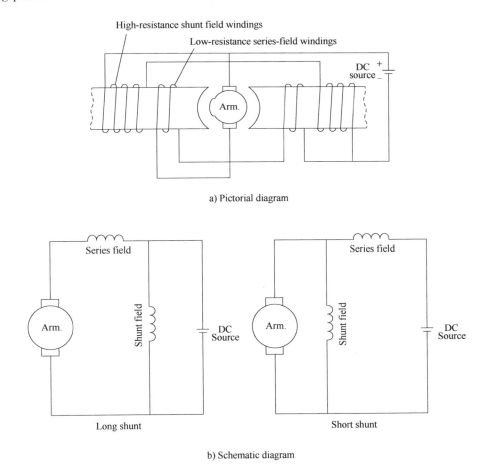

Fig. 1.12 Compound-wound DC motor

Compound motors have high torque similar to a series-wound motor, together with good speed regulation similar to a shunt motor. Therefore, when good torque and good speed regulation are needed, the compound wound dc motor can be used. A major disadvantage of a compound-wound motor is its expense.

2. **Types of AC Motors**

(1) Three-Phase Induction Motor

This is one of the most rugged and most widely used machines in industry. Its stator is composed of laminations of high-grade sheet steel. The inner surface is slotted to accommodate a three-phase winding. In Fig. 1.13 the three-phase winding is represented by three coils, the axes of which are 120° electrical degrees apart. Coil aa' represents all the coils assigned to phase a for one pair of poles. Similarly coil bb' represents phase b coils, and coil cc' represents phase c coils. When one end of each phase is tied together, the three phase stator winding is said to be Y-connected. Such a winding is called a three-phase winding because the voltages induced in each of the three phases by a revolving flux density field are out of phase by 120° electrical degrees a distinguishing characteristic of a balanced three-phase system.

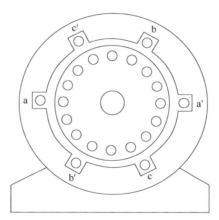

Fig. 1.13 Three-phase induction motor

The rotor also consists of laminations of slotted ferromagnetic material, but the rotor winding may be either the squirrel-cage type or the wound-rotor type. The latter is of a form similar to that of the stator winding. The winding terminals are brought out to three slip rings. This allows an external three-phase resistor to be connected to the rotor winding for the purpose of providing speed control. As a matter of fact, it is the need for speed control which in large measure accounts for the use of the wound-rotor type induction motor. Otherwise the squirrel-cage induction motor would be used. The squirrel-cage winding consists merely of a number of copper bars embedded in the rotor slots and connected at both ends by means of copper end rings. (In some of the smaller sizes aluminum is used.) The squirrel-cage construction is not only simpler and more economical than the wound-rotor type but more rugged as well. There are no slip rings or carbon brushes to be bothered with.

The revolving field produced by stator winding cuts the rotor conductors, thereby inducing voltages. Since the rotor winding is short-circuited by the end rings, the induced voltages cause currents to flow which in turn react with the field to produce electromagnetic torque, and so motor action results.

Accordingly, on the basis of the foregoing description, it should be clear that for the three-phase induction motor the field winding is located on the stator and the armature winding on the rotor. Another point worth noting is that this machine is singly excited, i. e., electrical power is applied only to the stator winding. Current flows through the rotor winding by induction. As a consequence both the magnetizing current, which sets up the magnetic field, and the power current, which allows energy to be delivered to the shaft load, flow through the stator winding. For this reason, and in the interest of keeping the magnetizing current as small as possible in order that the power component may be correspondingly larger for a given current rating, the air gap of induction motors is made as small as mechanical clearance will allow. The air-gap lengths vary from about 0. 02 in. for smaller machines to 0. 05 in. for machines of higher rating and speed.

(2) Synchronous Machines

The essential construction features of the synchronous machine are depicted in Fig. 1. 14. The stator consists of a stator frame, a slotted stator core, which provides a low-reluctance path for the magnetic flux, and a three-phase winding imbedded in the slots. The rotor either is cylindrical and equipped with a distributed winding or else has salient-poles with a coil wound on each leg as depicted in Fig. 1. 14. The cylindrical construction is used exclusively for turbo-generators, which operate at high speeds. On the other hand, the salient-pole construction is used exclusively for synchronous motors operating at speeds of 1800 r/min or less.

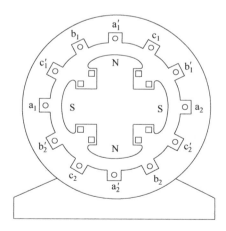

Fig. 1. 14   Salient-pole synchronous machine

When operated as a generator the synchronous machine receives mechanical energy from a prime mover such as a steam turbine and is driven at some fixed speed. Also, the rotor winding is energized from a DC source, thereby furnishing a field distribution along the air gap. When the rotor is at standstill and DC flows through the rotor winding, no voltage is induced in the stator winding because the flux is not cutting the stator coils. However, when the rotor is being driven at full speed, voltage is induced in the stator winding and upon application of a suitable load electrical energy may be delivered to it.

For the synchronous machine the field winding is located on the rotor, the armature winding is located on the stator. This statement is valid even when the synchronous machine operates as a motor. In this mode AC power is applied to the stator winding and DC power is applied to the rotor winding for the purpose of energizing the field poles. Mechanical energy is then taken from the shaft. Note, too, that unlike the induction motor, the synchronous motor is a doubly excited machine; i.e., energy is applied to the rotor as well as the stator winding. In fact it is this characteristic which enables this machine to develop a nonzero torque at only one speed—hence the name synchronous.

Because the magnetizing current for the synchronous machine originates from a separate source (the DC supply), the air-gap lengths are larger than those found in induction motors of comparable size and rating. However, synchronous machines are more expensive and less rugged than induction motors in the smaller horsepower ratings because the rotor must be equipped with slip rings and brushes in order to allow the direct current to be conducted to the field winding.

1.5.2 Speed Control of DC Motors

1. DC Motor Analysis

A DC motor is a DC generator with the power flow reversed. In the DC motor electrical energy is converted to mechanical form. On the basis of foregoing discussion, there are three types of DC motors: the shunt motor, the cumulatively compounded motor, and the series motor. The compound motor is prefixed with the cumulative in order to stress that the connections to the series field winding are such as to ensure that the series field flux aids the shunt-field flux. The series motor, unlike the series generator, finds wide application, especially for traction-type loads. Hence due attention is given to this machine in the treatment that follows.

The performance of the DC motor operating in any one of its three modes can conveniently be described in terms of an equivalent circuit, a set of performance equations, a power-flow diagram and the magnetization curve. The equivalent circuit is depicted in Fig. 1.15. It is worthwhile to note that now the armature induced voltage is treated as a reaction or counter emf.

By imposing some constrains, we obtain the correct equivalent circuit for the desired mode of operation.

For example, for a series motor the appropriate equivalent circuit results upon removing R_f from the circuitry of Fig. 1.15.

The set of equations needed to compute the performance is listed below:

$$E_a = K_E \Phi_n \quad (1\text{-}1)$$

$$T = K_T \Phi I_a \quad (1\text{-}2)$$

$$U_t = E_a + I_a(R_a + R_s) \quad (1\text{-}3)$$

$$I_L = I_f + I_a \quad (1\text{-}4)$$

Note that the last two equations are modified to account for the fact that for the motor U_t is the

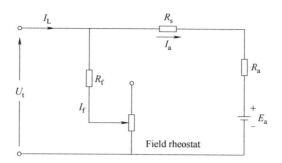

Fig. 1.15 Equivalent circuit of the DC motor

applied or source voltage and as such must be equal to the sum of the voltage drops. Similarly, the line current is equal to the sum rather than the difference of the armature current and field currents.

The power-flow diagram is illustrated in Fig. 1.16. The electrical power input $U_t I_L$ originating from the line supplies the field power needed to establish the flux field as well as the armature-circuit copper loss needed to maintain the flow of I_a. This current flowing through the armature conductors imbedded in the flux field causes torque to be developed. The law of conservation of energy then demands that the electromagnetic power $E_a I_a$ be equal to $T\omega_m$, where ω_m is the steady-state operating speed. Removal of the rotational losses from the developed mechanical power yields the mechanical output power.

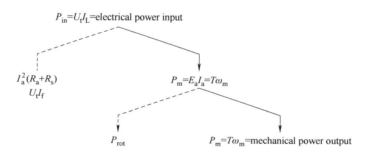

Fig. 1.16 Power-flow diagram of DC motor

The DC motor is often called upon to do the really tough jobs in industry because of its high degree of flexibility and ease of control. These features cannot easily be matched by other electromechanical energy-conversion devices. The DC motor offers a wide range of control of speed and torque as well as excellent acceleration and deceleration. For example, by the insertion of an appropriate armature-circuit resistance, rated torque can be obtained at starting with no more than rated current flowing. Also, by special design of the shunt-field winding, speed adjustments over a range of 4∶1 are readily obtainable. If this is then combined with armature-voltage control, the range of speed adjustment spreads to 6∶1. In some electronic control devices that are used to provide the DC energy to the field and armature circuits, a speed range of 40∶1 is possible. The size of the motor being controlled, however, is limited.

2. DC Motor Speed-Torque Characteristics

How does the DC motor react to the application of a shaft load? What is the mechanism by which the DC motor adapts itself to supply to the load the power it demands? The answer to these questions can be obtained by reasoning in terms of the performance equations. Initially our remarks are confined to the shunt motor, but a similar line of reasoning applies for the others. For our purposes the two pertinent equations are those for torque and current. Thus

$$T = K_T \Phi I_a \tag{1-5}$$

$$I_a = \frac{U_t - K_E \Phi n}{R_a} \tag{1-6}$$

Note that the last expression results from replacing E_a by Equation (1-1) in Equation (1-3). With no shaft load applied, the only torque needed is that which overcomes the rotational losses. Since the shunt motor operates at essentially constant flux, Equation (1-2) indicates that only a small armature current is required compared to its rated value to furnish these losses. Equation (1-6) reveals the manner in which the armature current is made to assume just the right value. In this expression U_t, R_a, K_E, and Φ are fixed in value. Therefore the speed is the critical variable. If, for the moment, it is assumed that the speed has too low a value, then the numerator of Equation (1-6) takes on an excessive value and in turn makes I_a larger than required. At this point the motor reacts to correct the situation. The excessive armature current produces a developed torque, which exceeds the opposing torques of friction and windage. In fact this excess serves as an accelerating torque, which then proceeds to increase the speed to that level which corresponds to the equilibrium value of armature current. In other words, the acceleration torque becomes zero only when the speed is at that value which by Equation (1-6) yields just the right I_a needed to overcome the rotational losses.

Consider next that a load demanding rated toque is suddenly applied to the motor shaft. Clearly, because the developed torque at this instant is only sufficient to overcome friction and windage and not the load torque, the first reaction is for the motor to lose speed. In this way, as Equation (1-6) reveals, the armature current can be increased so that in turn the electromagnetic torque can increase.

As a matter of fact the applied load torque causes the motor assume that value of speed which yields a current sufficient to produce a developed torque to overcome the applied shaft torque and the frictional torque. Power balance is thereby achieved, because an equilibrium condition is reached where the electromagnetic power, $E_a I_a$, is equal to the mechanical power developed, $T\omega_m$.

A comparison of the DC motor with the three-phase induction motor indicates that both are speed-sensitive devices in response to applied shaft loads. An essential difference, however, is that for the three-phase induction motor developed torque is adversely influenced by the power-factor angle of the armature current. Of course no analogous situation prevails in the case of the DC motor.

On the basis of he foregoing discussion it should be apparent that the speed-torque curve of DC motors is an important characteristic. Appearing in Fig. 1.17 are the general shapes of the speed-torque characteristics as they apply for the shunt, cumulatively compounded, and series motors. For the sake of comparison the curves are drawn through a common point of rated torque and speed. An understanding of why the curves take the shapes and relative positions depicted in Fig. 1.17 readily follows from an examination of Equation (1-1), which involves the speed. For the shunt motor the speed equation can be written as

$$n = \frac{E_a}{K_E \Phi_{sh}} = \frac{U_t - I_a R_a}{K_E \Phi_{sh}} \tag{1-7}$$

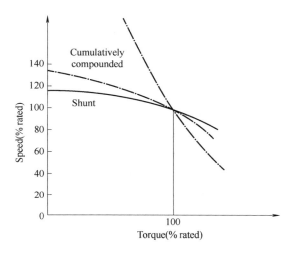

Fig. 1.17 Typical speed-torque curves of DC motors

The only variables involved are the speed n and the armature current I_a. At rated output torque the armature current is at its rated value and so, too, is the speed. As the load torque is removed, the armature current becomes correspondingly smaller, making the numerator term of Equation (1-7) larger. This results in higher speeds. The extent to which the speed increases depends upon how large the armature circuit resistance drop is in comparison to the terminal voltage. It is usually around 5%-10%. Accordingly, we can expect the percent change in speed of the shunt motor to be about the same magnitude. This change in speed is identified by a figure of merit called the speed regulation. It is defined as follows

$$\text{percent speed regulation} = \frac{(\text{no load speed}) - (\text{full load speed})}{\text{full load speed}} \tag{1-8}$$

The speed equation as it applies to the cumulatively compounded motor takes the form

$$n = \frac{U_t - I_a(R_a + R_s)}{K_E(\Phi_{sh} + \Phi_s)} \tag{1-9}$$

A comparison with analogous expression for the shunt motor bears out two differences. One, the numerator term also includes the voltage drop in the series-field winding besides that in the armature winding. Two, the denominator term is increases to account for the effect of the series-field flux Φ_s. Starting at rated torque and speed, Equation (1-8) makes it clear that as load torque is decreased to zero there is an increase in the numerator term which is necessarily greater than it is for the shunt motor. At the same time, moreover, the denominator term decreases because Φ_s reduces to zero as the torque goes to zero. Both effects act to bring about an increase in speed. Therefore, the speed regulation of the cumulatively compounded motor is greater than for the shunt motor. Fig. 1.17 presents this information graphically.

The situation regarding the speed-torque characteristic of the series motor is significantly different because of the absence of a shunt-field winding. Keep in mind that the establishment of a flux field in the series motor comes about solely as a result of the flow of armature current through the series-field winding. In this connection, then, the speed equation for the series motor becomes

$$n = \frac{E_a}{K_E \Phi_s} = \frac{U_t - I_a(R_a + R_s)}{K'_E I_a} \tag{1-10}$$

where K'_E denotes a new proportionality factor which permits Φ_s to be replaced by the armature current I_a. When rated torque is being developed the current is at its rated value. The flux field is therefore abundant. However, as load torque is removed less armature current flows. Now since I_a appears in the denominator of the speed equation, it is easy to see that the speed will increase greatly. In fact, if the load were to be disconnected from the motor shaft, dangerously high speeds would result because of the small armature current that flows. The centrifugal forces at these high speeds can easily damage the armature winding. For this reason a series motor should never have its load uncoupled.

Because the armature current is directly related to the air-gap flux in the series motor, Equation (1-1) for the developed torque may be modified to read as

$$T = K_T \Phi I_a = K'_T I_a^2 \tag{1-11}$$

Thus the developed torque for the series motor is a function of the square of the armature current. This stands in contrast to the linear relationship of torque to armature current in the shunt motor. Of course in the compound motor an intermediate relationship is achieved. It is interesting to note, too, that as the series motor reacts to develop greater torques, the speed drops correspondingly. It is this capability which suits the series motor so well to traction-type loads.

One of the attractive features the DC motor offers over all other types is the relative ease with which speed control can be achieved. The various schemes available for speed control can be deduced from Equation (1-7), which is repeated here with one modification

$$n = \frac{U_t - I_a(R_a + R_s)}{K_E \Phi} \tag{1-12}$$

The modification involves the inclusion of an external armature-circuit resistance R_e. Inspection of Equation (1-12) reveals that the speed can be controlled by adjusting any one of the three factors appearing on the right side of the equation: U_t, R_e, or Φ.

The simplest is adjust Φ. A field rheostat such as that shown in Fig. 1.15 is used. If the field-rheostat resistance is increased, the air-gap flux is diminished, yielding higher operating speeds. General-purpose shunt motors are designed to provide a 200% increase in rated speed by this method of speed control. However, because of the weakened flux field the permissible torque that can be delivered at the higher speed is correspondingly reduced, in order to prevent excessive armature current.

A second method of speed adjustment involves the use of an external resistor R_e connected in the armature circuit as illustrated in Fig. 1.18. The size and cost of this resistor are considerably greater than those of the field rheostat because R_e must be capable of handling the full armature current. Equation (1-12) indicates that the larger R_e is made, the greater will be the speed change. Frequently the external resistor is selected to furnish as much as a 50% drop in speed from the rated value. The chief disadvantage of this method of control is the poor efficiency of operation. For example, a 50% drop in speed is achieved by having approximately half of the terminal voltage U_t appear across R_e. Accordingly, almost 50% of the line input power is dissipated in the form of heat in the resistor R_e. Nonetheless, armature-circuit resistance control is often used—especially for series motors.

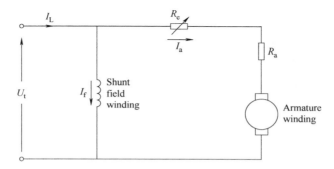

Fig. 1.18 Speed adjustment of a shunt motor by an external armature-circuit resistance

A third and final method of speed control involves adjustment of the applied terminal voltage. This scheme is the most desirable from the viewpoint of flexibility and high operating efficiency. But it is also the most expensive because it requires its own DC supply. It means purchasing a motor-generator set with a capacity at least equal to that of the motor to be controlled. Such expense is not generally justified except in situations where the superior performance achievable with this scheme is indispensable, as is the case in steel mill applications. Armature terminal voltage control is referred to as the Ward-Leonard system.

3. Electrical Braking

In many speed control systems, e.g., rolling mills, mine winders, etc., the load has to be frequently brought to a standstill and reversed. The rate at which the speed reduces following a reduced speed demand is dependent on the stored energy and the braking system used. A small speed control system (sometimes known as a velodyne) can employ mechanical braking, but this is not feasible with large speed controllers since it is difficult and costly to remove the heat generated.

The various methods of electrical braking available are:
1) Regenerative braking.
2) Eddy current braking.
3) Dynamic braking.
4) Reverse current braking (plugging).

Regenerative braking is the best method, though not necessarily the most economic. The stored energy in the load is converted into electrical energy by the work motor (acting temporarily as a generator) and is returned to the power supply system. The supply system thus acts as a "sink" into which the unwanted energy is delivered. Providing the supply system has adequate capacity, the consequent rise in terminal voltage will be small during the short periods of regeneration. In the Ward-Leonard method of speed control of DC motors, regenerative braking is inherent, but thyristor drives have to be arranged to invert to regenerate. Induction motor drives can regenerate if the rotor shaft is driven faster than the speed of the rotating field. The advent of low-cost variable-frequency supplies from thyristor inverters have brought about considerable changes in the use of induction motors in variable speed drives.

Eddy current braking can be applied to any machine, simply by mounting a copper or aluminum disc on the shaft and rotating it in a magnetic field. The problem of removing the heat generated is severe in large systems as the temperature of the shaft, bearings, and motor will be raised if prolonged braking is applied.

In dynamic braking, the stored energy is dissipated in a resistor in the circuit. When applied to small DC machines, the armature supply is disconnected and a resistor is connected across the armature (usually by a relay, contactor, or thyristor). The field voltage is maintained, and braking is applied down to the lowest speeds. Induction motors require a somewhat more complex arrangement, the stator windings being disconnected from the AC supply and reconnected to a DC supply. The electrical energy generated is then dissipated in the rotor circuit. Dynamic braking is applied to many large AC hoist systems where the braking duty is both severe and prolonged.

Any electrical motors can be brought to a standstill by suddenly reconnecting the supply to reverse the direction of rotation (reverse current braking). Applied under controlled conditions, this method of braking is satisfactory for all drives. Its major disadvantage is that the electrical energy consumed by the machine when braking is equal to the stored energy in the load. This increases the running costs significantly in large drives.

4. A Single-Quadrant Speed Control System Using Thyristors

A single-quadrant thyristor converter system is shown in Fig. 1.19. For the moment the reader should ignore the rectifier BR_2 and its associated circuitry.

Since the circuit is a single-quadrant converter, the speed of the motor shaft (which is the output from the system) can be controlled in one direction of rotation only. Moreover, regenerative braking can not be applied to the motor; in this type of system, the motor armature can suddenly be brought to rest by dynamic braking (i.e. when the thyristor gate pulses are phased back to 180°, a resister can be connected across the armature by a relay or some other means).

Rectifier BR_1 provides a constant voltage across the shunt field winding, giving a constant field flux. The armature current is controlled by a thyristor which is, in turn, controlled by the pulses applied to its gate. The armature speed increases as the pulses are phased forward (which reduces the delay angle of firing), and the armature speed reduces as the gate pulses are phased back.

The speed reference signal is derived from a manually operated potentiometer (shown at the right-hand side of Fig. 1.20), and the feedback signal or output speed signal is derived from the

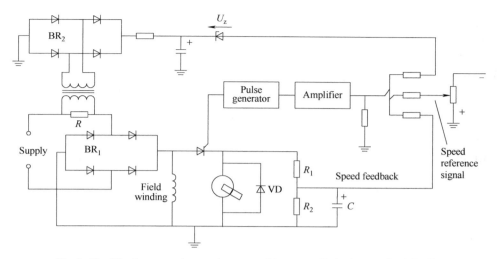

Fig. 1.19　Thyristor speed control system with current limitation on the AC side

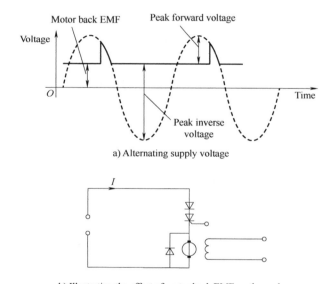

a) Alternating supply voltage

b) Illustrating the effect of motor back EMF on the peak inverse voltage applied to the thyristor

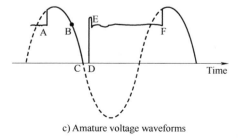

c) Amature voltage waveforms

Fig. 1.20　Thyristor speed control system with current limitation on the AC

resistor chain $R_1 R_2$, which is connected across the armature. Since the armature voltage is obtained from a thyristor, the voltage consists of a series of pulses, these pulses are smoothed by capacitor C. The speed reference signal is of the opposite polarity to the armature voltage signal to ensure that overall negative feedback is applied.

A feature of DC motor drives is that the load presented to the supply is a mixture of resistance, inductance, and back EMF diode VD in Fig. 1. 19. ensures that the thyristor current commutates to zero when its anode potential falls below the potential of the upper armature connection, in the manner outlined before. In the drive shown, the potential of the thyristor cathode is equal to the back EMF of the motor while it is in a blocking state.

1.5.3 Speed Control of AC Motors

1. Three-Phase Induction Motor

One distinguishing feature of the induction motor is that it is a singly excited machine. Although such machines are equipped with both a field winding and an armature winding, in normal use an energy source is connected to one winding alone, the field winding. Currents are made to flow in the armature winding by induction, which creates an ampere-conductor distribution that interacts with the field distribution to produce a net unidirectional torque. The frequency of the induced current in the conductor is affected by the speed of the rotor on which it is located; however, the relationship between the rotor speed and the frequency of the armature current is such as to yield a resulting ampere-conductor distribution that is stationary relative to the field distribution of the stator. As a result, the singly excited induction machine is capable of producing torque at any speed below synchronous speed. For this reason the induction machine is placed in the class of asynchronous machines. In contrast, synchronous machines are electromechanical energy-conversion devices in which a net torque can be produced at only one speed of the rotor. The distinguishing characteristic of the synchronous machine is that it is a doubly excited device except when it is being used as a reluctance motor.

Because the induction machine is singly excited, it is necessary that the magnetizing current and the power component of the current flow in the same lines. Moreover, because of the presence of an air gap in the magnetic circuit of the induction machine, an appreciable amount of magnetizing current is needed to establish the flux per pole demanded by the applied voltage. Usually, the value of the magnetizing current for three-phase induction motors lies between 25% and 40% of the rated current. Consequently, the induction motor is found to operate at a low power factor at light loads and at less than unity power factor in the vicinity of rated output.

The primary functions of a controller are to furnish proper starting, stopping and reversing without damage or inconvenience to the motor, other connected loads, or the power system. However, the controller fulfills other useful purposes as well, especially the following:

1) It limits the starting torque. Some connected shaft loads may be damaged if excessive torque is applied upon starting. For example, fan blades can be sheared off or gears having excessive backlash can be stripped. The controller supplies reduced voltage at the start and as the speed picks up the voltage is increased in steps to its full value.

2) It limits the starting current. Most motors above 2.38 kW cannot be started directly across the three-phase line because of the excessive starting current that flows. Recall that at unity slip the current is limited only by the leakage impedance, which is usually quite a small quantity, especially in the larger motor sizes. A large starting current can be annoying because it causes light to flicker and may even cause other connected motors to stall. Reduced voltage starting readily eliminates these annoyances.

3) It provides overload protection. All general-purpose motors are designed to deliver full-load power continuously without overheating. However, if for some reasons the motor is made to deliver, say, 150% of its rated output continuously, it will proceed to accommodate the demand and burn itself up in the process. The horsepower rating of the motor is based on the allowable temperature rise that can be tolerated by the insulation used for the field and armature windings. The losses produce the heat that raises the temperature. As long as these losses do not exceed the rated values there is no danger to the motor, but if they are allowed to become excessive, damage will result. There is nothing inherent in the motor that will keep the temperature rise within safe limits. Accordingly, it is also the function of the controller to provide this protection. Overload protection is achieved by the use of an appropriate time-delay relay which is sensitive to the heat produced by the motor line currents.

4) It furnishes under-voltage protection. Operation at reduced voltage can be harmful to the motor, especially when the loads demand rated power. If the line voltage falls below some preset limits, the motor is automatically disconnected from the three-phase source by the controller.

2. **Torque-Speed Characteristics of Induction Motor**

When the motor operates at a very small slip, as at no-load, the rotor current is practically zero so that only that amount of torque is developed which is needed to supply the rotational losses. The slip is allowed to increase from nearly zero to about 10%.

As the slip is allowed to increase still further, the current continues to increase but much less rapidly than at first. The reason lies in the increasing importance of the reactance of the rotor impedance. In addition, the space angle now begins to increase at a rapid rate which makes it diminish more rapidly than the current increases. Since the torque equation now involves two opposing factors, it is entirely reasonable to expect that a point is reached beyond which further increases in slip culminate in decreased developed torque. In other words, the rapidly decreasing $\cos b$ factor, where b is the angle between the rotor induced electromotive force and the current which are reduced to the stator side, predominates over the slightly increasing rotor current which is reduced to the stator side. As it increases, the field pattern for producing torque becomes less and less favorable because more and more conductors that produce negative torque are included beneath a given pole flux.

The starting torque is the torque developed when s is unity, i.e. the speed n is zero. Fig. 1.21 indicates that for the case illustrated the starting torque is somewhat in excess of speed rated torque, which is fairly typical of such machines. The starting torque is computed in three-phase induction motor the same manner as torque is computed for any value of slip. Here it merely requires using $s = 1$.

It is interesting to note that higher starting torques result from increased rotor copper losses at

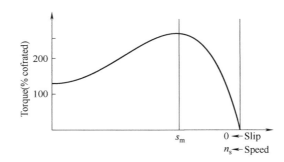

Fig. 1.21 Typical torque-speed curve of a starting torque

standstill.

3. Adjustable-Frequency Concepts

Let's review for a moment the concepts of line and forced commutation as they are used to obtain adjustable frequency to be applied to an AC squirrel cage induction motor. Fig. 1.22 illustrates what the controller is supposed to do: create an adjustable voltage and adjustable frequency from fixed line voltage and fixed line frequency. Let's first decide what we need to put in the box.

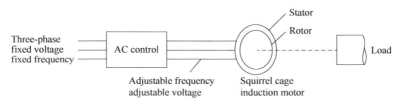

Fig. 1.22 Conversion functions of adjustable frequency controllers applied to AC motors

The circuit using six thyristors will not work. It can create an adjustable voltage to the motor, but the line frequency passes straight on through. Therefore, we need a means of creating an adjustable frequency as well as an adjustable voltage. The simplest way of doing this is by means of a DC link. The DC link is then controlled by one of several means to create the adjustable frequency. In some cases the DC link is also controlled to create the adjustable voltage. To form the DC link, the incoming AC voltage must somehow be changed to a DC voltage, after which the DC is changed back to AC for applying to the AC motor.

Fig. 1.23 shows a generalized frequency controller with a DC link. The input uses six semiconductors to provide the DC, and the output uses six semiconductors to provide the adjustable frequency. Which type of semiconductor should be selected for each box? Since the AC line is always connected to the first box, the input devices can be line commutated. They can therefore be diodes, thyristors, GTOs, transistors, or triacs. GTOs are quite costly, and transistors and triacs may not have the desired ampere capacity and voltage capabilities. Therefore, the input power devices will be diodes or thyristors or perhaps a combination.

Let's start with diodes in the input box. With diodes, there is no means of adjusting the DC link voltage. Therefore, both the adjustable frequency and the adjustable voltage must be created in the output stage. This is actually done in the real case. The resulting system is called pulse-width

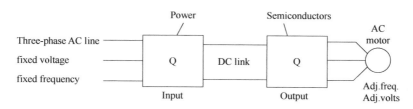

Fig. 1.23 Controlled adjustable frequency controller with DC Link

modulation (PWM).

If the input box were to use thyristors instead of diodes, they can be controlled to provide adjustable DC link voltage. The output stage then needs only to create the adjustable frequency from the DC link and pass the adjustable voltage on through to the AC motor, together with the adjustable frequency. The output stage can therefore be less complex than for a PWM system, but the input stage must be more complex. However, this method is also in popular use for frequency controllers, in one of two arrangements. One is known as the adjustable voltage inverter (AVI). The other is known as the current source inverter (CSI).

VOCABULARY

1. electricity generation   发电
2. electricity transmission   输电
3. reciprocating engine   往复式引擎
4. permanent magnet   永久磁铁
5. generator   *n.* 发电机
6. photoelectric   *adj.* 光电的
7. iron core   铁心
8. turbine   *n.* 涡轮机
9. thermal energy   热能
10. condense   *v.* 使凝结
11. component   *n.* 元器件
12. turbine-generator   涡轮发电机
13. hydroelectric   *adj.* 水力发电的
14. excitation   *n.* 励磁
15. synchronous   *adj.* 同步的
16. overload   *n.* 过载
17. exciter   *n.* 励磁机
18. regulator   *n.* 调节器
19. amplidyne   *n.* 电机放大机
20. alternator   *n.* 交流发电机
21. separately excited   他励的
22. armature   *n.* 电枢
23. air-gap   气隙
24. servo-control   伺服控制
25. field winding   励磁绕组
26. magnetic field   磁场
27. polarity   *n.* 极性
28. commutator   *n.* 换相器
29. series-wound   串励
30. shunt-wound   并励
31. compound-wound   复励
32. brush   *n.* 电刷
33. horsepower   *n.* 马力
34. demagnetization   *n.* 去磁, 失磁
35. rheostat   *n.* 变阻器
36. field current   励磁电流
37. cumulative compound   积式复励
38. differential compound   差分复励
39. short shunt   短复励
40. long-shunt   长复励
41. ferromagnetic material   铁磁材料
42. squirrel-cage   笼型

43. wound-rotor 绕线转子
44. stator winding 定子绕组
45. three-phase induction motor 三相感应电机
46. low-reluctance 低磁阻
47. magnetic flux 磁通
48. doubly excited 双励
49. slip ring 集电环
50. regenerative braking 回馈制动
51. eddy current braking 涡流制动
52. dynamic braking 能耗制动
53. reverse current braking 电流反向接制动
54. variable-frequency 变频
55. copper *n.* 铜
56. bearing *n.* 轴承
57. field voltage 励磁电压
58. electrical motor 电机
59. single-quadrant 单象限
60. field flux 励磁磁通
61. reference signal 参考信号、给定信号
62. power factor 功率因数
63. starting current 起动电流
64. overload protection 过载保护
65. time-delay relay 时间继电器
66. under-voltage protection 欠电压保护
67. slip *n.* 转差
68. adjustable-frequency 调频
69. DC link 直流环节；直流母线

NOTES

1. Electricity is most often generated at a power station by electromechanical generators, primarily driven by heat engines fueled by chemical combustion or nuclear fission but also by other means such as the kinetic energy of flowing water and wind.

电能通常由发电厂中机电能量转换的发电机产生。发电机主要由采用化学燃烧或核裂变产热的热机驱动，但也有利用水和风的动能等其他方式驱动。

2. The combustion of fossil fuels supplies most of the heat to these engines, with a significant fraction from nuclear fission and some from renewable sources.

这些发动机需要的热量绝大多数由化石燃料燃烧提供，还有很大一部分来自于核裂变和可再生能源。

3. This current flows in one direction as the magnet's north pole approaches a coil and in the opposite direction as the south pole approaches it. To generate the 60 Hz alternating current used in the United States, the generator must turn 60 times each second so that the current completes one full cycle of reversals every 1/60th of a second.

当磁铁的 N 极接近线圈时电流沿一个方向流动，当磁铁的 S 极接近线圈时电流沿相反的方向流动。在美国，为了产生 60 Hz 的交流电，发电机必须每秒转 60 圈，这样电流每 1/60 s 内完成一个周期。

4. The basic requirement is that the excitation system supply and automatically adjust the field current of the synchronous generator to maintain the terminal voltage as the output power varies within the capability of the generator.

最基本的要求是励磁系统为同步发电机提供并自动调节励磁电流，当输出功率在发电机容量范围内变化时，维持端电压恒定。

5. The excitation systems of this category utilize DC generators as sources of excitation power

and provide the field current to the rotor of the synchronous machine through slip rings.

这种励磁系统利用直流发电机作为励磁电源，通过集电环为同步发电机转子提供励磁电流。

6. The use of electrical motors has increased for home appliances and industrial and commercial applications for driving machines and sophisticated equipment.

电动机在家用电器以及工业和商用的驱动电动机和复杂装置方面的应用更广泛了。

7. Each of the types of DC motors are similar in construction to the type of DC generator that corresponds to it. The only difference, in most cases, is that the generator acts as a voltage source while the motor functions as a mechanical power conversion device.

每种类型的直流电动机和相应的直流发电机的结构是相似的。大多数情况下，唯一的区别在于发电机作为一个电压源使用，而电动机作为一个机电转换设备使用。

8. The speed of a series motor varies from very fast at no load to very slow at heavy loads. Since large currents may flow through the low-resistance field, the series motor produces a high-torque output.

串励电动机从空载到重载时转速变化很快。由于很大的电流流过低电阻的励磁线圈，串励电动机可以产生很大的输出转矩。

9. The revolving field produced by stator winding cuts the rotor conductors, thereby inducing voltages. Since the rotor winding is short-circuited by the end rings, the induced voltages cause currents to flow which in turn react with the field to produce electromagnetic torque, and so motor action results.

定子线圈所产生的旋转磁场切割转子导体，进而产生感应电压。由于转子绕组通过末端绕组短接，感应电压便会引起电流流通进而与磁场作用产生电磁转矩，这样电动机就开始运转起来。

10. Because the magnetizing current for the synchronous machine originates from a separate source (the DC supply), the air-gap lengths are larger than those found in induction motors of comparable size and rating.

因为同步电动机的励磁电流是由独立电源（直流电）产生，气隙长度要比近似尺寸和容量等级的异步电动机更大一些。

11. As a result, the singly excited induction machine is capable of producing torque at any speed below synchronous speed. For this reason the induction machine is placed in the class of asynchronous machines.

由于单励感应电动机可以在低于同步转速时产生有效转矩。因而，感应电动机归为异步电动机一类。

12. However, if for some reasons the motor is made to deliver, say, 150% of its rated output continuously, it will proceed to accommodate the demand and bum itself up in the process. The horsepower rating of the motor is based on the allowable temperature rise that can be tolerated by the insulation used for the field and armature windings.

但是，如果由于某些原因电动机需要连续超载传输，如150%的额定输出，电动机会继续负载运行然后被烧坏。电动机的额定功率值是基于励磁绕组和电枢绕组的绝缘材料所允许的温升来设定的。

Chapter 2
Transmission System

2.1 Introduction

Electrical power is transformerred from generating stations to consumers through overhead lines and cables.

Overhead lines are used for long distances in open country and rural areas. Whereas cables are used for underground transmission in urban areas and for underwater crossings. For the same rating, cables are 10 to 15 times more expensive than overhead lines and they are therefore only used in special situations where overhead lines cannot be used; the distances in such application are short.

2.2 Transmission Lines

2.2.1 Overhead Lines

A transmission line is characterized by four parameters: series resistance R due to the conductor resistivety, shunt conductance G due to leakage currents between the phase and ground, series inductance L due to magnetic field surrounding the conductor, and shunt capacitance C due to the electric field between conductors.

Detailed derivations from first principles for the line parameters can be found in standard books on power systems. Here, we will briefly summarize salient points relating to line parameters.

1) Series resistance R. The resistances of lines accounting for standing and skin effect are determined from manufacturers' tables.

2) Shunt conductance G. The shunt conductance represents losses due to leakage currents along insulator strings and corona. In power lines, its effect is small and usually neglected.

3) Series inductance L. The line inductance depends on the partial flux linkages within the conductor cross section and external flux linkages. For overhead lines, the inductances of the three phases are different from each other unless the conductors have equilateral spacing, a geometry not usually adopted in practice. The inductances of the three phases with non-equilateral spacing can be equalized by transposing the lines in such a way that each phase occupies successively all three possible positions.

For a transposed three-phase line, the inductance per phase is

$$L = 2 \times 10^{-7} \ln \frac{D_{eq}}{D_s} \text{ H/m} \tag{2-1}$$

In the above equation, D_s is the self geometric mean distance, taking into account the conductor composition, stranding, and bundling; it is also called the geometric mean radius. And D_{eq} is the geometric mean of the distances between the conductors of the three-phase a, b, and c, it is expressed as

$$D_{eq} = (d_{ab} d_{bc} d_{ca})^{1/3} \tag{2-2}$$

4) Shunt capacitance C. The potential difference between the conductors of a transmission line causes the conductors to be charged; the charge per unit of potential difference is the capacitance between conductors when alternating voltages are applied to the capacitances.

2.2.2 Underground Cables

Underground cables have the same basic parameters as overhead lines: series resistance and inductance; shunt capacitance and conductance.

However, the values of the parameters and hence the characteristic of cables differ significantly from those of overhead lines for the following reasons:

1) The conductors in a cable are much closer to each other than the conductors of overhead lines.

2) The conductors in a cable are surrounded by metallic bodies such as shield, lead or aluminum sheets, and steel pipes.

3) The insulating material between conductors in a cable is usually impregnated paper, low-viscosity oil, or an inert gas.

In the previous section, we identified the parameters of a transmission line per unit length. These are distributed parameters; that is, the effects represented by the parameters are distributed throughout the length of the line.

If the line is assumed transposed, we can analyze the line performance on a per-phase basis. Fig. 2.1 shows the relationship between current and voltage along one phase of the line in terms of the distributed parameters, with

$z = R + j\omega L$ = series impedance per unit length/phase

$y = G + j\omega C$ = shunt admittance per unit length/phase

l = length of the line

The voltages and currents shown are phasors representing sinusoidal time-varying quantities.

Consider a differential section of the line of length dx at a distance x from the receiving end. The differential voltage across the elemental length is given by

$$d\tilde{V} = \tilde{I}(z dx)$$

Hence

$$\frac{d\tilde{V}}{dx} = \tilde{I}z \tag{2-3}$$

The differential current flowing into the shunt admittance is

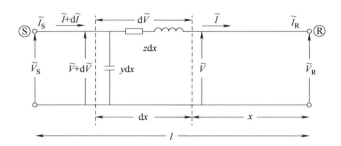

Fig. 2.1 Voltage and current relationship of a distributed parameter line

$$d\tilde{I} = \tilde{V}(y dx)$$

Hence

$$\frac{d\tilde{I}}{dx} = \tilde{V}y \qquad (2\text{-}4)$$

Differentiating Equations (2-3) and (2-4) with respect to x, we obtain

$$\frac{d^2\tilde{V}}{dx^2} = z\frac{d\tilde{I}}{dx} = yz\tilde{V} \qquad (2\text{-}5)$$

$$\frac{d^2\tilde{I}}{dx^2} = y\frac{d\tilde{V}}{dx} = yz\tilde{I} \qquad (2\text{-}6)$$

We will establish the boundary conditions by assuming that voltage V_R and current I_R are known at the receiving end ($x = 0$). The general solution of Equations (2-5) and (2-6) for voltage and current at a distance x from the receiving end is

$$\tilde{V} = \frac{\tilde{V}_R + Z_c \tilde{I}_R}{2}e^{\gamma x} + \frac{\tilde{V}_R - Z_c \tilde{I}_R}{2}e^{-\gamma x} \qquad (2\text{-}7)$$

$$\tilde{I} = \frac{\tilde{V}_R/Z_c + \tilde{I}_R}{2}e^{\gamma x} - \frac{\tilde{V}_R/Z_c - \tilde{I}_R}{2}e^{-\gamma x} \qquad (2\text{-}8)$$

Where

$$z_c = \sqrt{z/y} \qquad (2\text{-}9)$$

$$\gamma = \sqrt{yz} = \alpha + j\beta \qquad (2\text{-}10)$$

The constant Z_c is called the characteristic impedance and γ is called the propagation constant.

The constants γ and Z_c are complex quantities. The real part of the propagation constant γ is called the attenuation constant α, and the imaginary part the phase constant β. Thus the exponential term $e^{\gamma x}$ may be expressed as

$$e^{\gamma x} = e^{(\alpha + j\beta)x} = e^{\alpha x}(\cos\beta x + j\sin\beta x) \qquad (2\text{-}11)$$

Therefore, the first term in Equation (2-7) increases in magnitude and advances in phase as the distance from the receiving end increases. This term is called the incident voltage.

The expanded form of the second exponential term is

$$e^{-\gamma x} = e^{-(\alpha + j\beta)x} = e^{-\alpha x}(\cos\beta x - j\sin\beta x) \qquad (2\text{-}12)$$

2.3 Transformers

2.3.1 Introduction

A transformer is a device used to convert voltage levels in an AC circuit. They have numerous uses in power systems. To begin, it is more efficient to transmit power at high voltages and low current than low voltage and high current. Conversely, lower voltages are safer and more economic for end use. Thus, transformers are used to step-up voltages from the generators and then used to step-down the voltage for end use. Another wide use of transformers is for instrumentation so that sensitive equipment can be isolated from the high voltages and currents of the transmission system. Transformers may also be used as means of controlling real power flow by phase-shifting.

Transformers enable utilization of different voltage levels across the system. From the viewpoints of efficiency and power-transfer capability, the transmission voltages have to be high, but it is not practically feasible to generate and consume power at these voltages. In modern electric power systems, the transmitted power undergoes four to five voltage transformations between the generators and the ultimate consumers. Consequently, the total MVA rating of all the transformers in a power system is about five times the total MVA rating of all the generators.

In addition to voltage transformation, transformers are often used for control of voltage and reactive power flow. Therefore, practically all transformers used for bulk power transmission and many distribution transformers have taps in one or more windings for changing the turns ratio. From the power system viewpoint, changing the ratio of transformation is required to compensate for variations in system voltages. Two types of tap-changing facilities are provided: off-load tap changing and under-load tap changing (ULTC). The off-load tap changing facilities require the transformer to be de-energized for tap changing; they are used when the ratio will need to be changed only to meet long-term variations due to load growth, system expansion, or seasonal changes. The ULTC is used when the changes in ratio need to be frequent; for example, to take care of daily variations in system conditions. The taps normally allow the ratio to vary in the range of ±10% to ±15%.

Transformers may be either three-phase units or three single-phase units. The latter type of construction is normally used for large extra-high voltage (EHV) transformers and for distribution transformers. Large EHV transformers are of single-phase design due to the cost of spare, insulation requirements, and shipping considerations. The distribution systems serve single-phase loads and are supplied by single-phase transformers.

When the voltage transformation ratio is small, autotransformers are normally used. The primary and secondary windings of autotransformers are interconnected so that the power to be transformed by magnetic coupling is only a portion of the total power transmitted through the transformer. There is thus inherent metallic connection between the primary side and secondary side circuits; this is unlike the conventional two-winding transformer which isolates the two circuits.

Autotransformers are usually Y connected, with neutrals solidly grounded to minimize the propagation of disturbances occurring on one side into the other side. It is a common practice to add a low-capacity delta-connected tertiary winding. The tertiary winding provides a path for third harmonic currents, thereby reducing their flow on the network. It also assists in stabilizing the neutral. Reactive compensation is often provided through use of switched reactors and capacitors on a tertiary bus.

As compared to the conventional two-winding transformer, the autotransformer has advantages of lower cost, higher efficiency, and better regulation. These advantages become less significant as the transformation ratio increases; hence, autotransformers are used for low transformation ratios (for example, 500/230 kV).

In interconnected systems, it sometimes becomes necessary to make electrical connections that form loop circuits through one or more power systems. To control the circulation of power and prevent overloading certain lines, it is usually necessary in such situations to use phase-angle transformers. Often it is necessary to vary the extent of phase shift to suit changing system conditions; this requires provision of on-load phase-shifting capability. Voltage transformation may also be required in addition to phase shift.

2.3.2 Representation of Two-Winding Transformers

1. Basic Equivalent Circuit in Physical Units

The basic equivalent circuit of a two-winding transformer with all quantities in physical units is shown in Fig. 2.2. The subscripts p and s refer to primary and secondary quantities, respectively.

The magnetizing reactance X_{mp} is very large and is usually neglected. For special studies requiring representation of transformer saturation, the magnetizing reactance representation may be approximated by moving it to the primary or the secondary terminals and treating it as a voltage-dependent variable shunt reactance.

2. Per Unit Equivalent Circuit

With appropriate choice of primary and secondary side base quantities, the equivalent circuit can be simplified by eliminating the ideal transformer. However, this is not always possible and the base quantities often have to be chosen independent of the actual turns ratio. It is therefore necessary to consider an off-nominal turns ratio.

From the equivalent circuit of Fig. 2.2, with X_{mp} neglected, we have

$$\tilde{v}_p = Z_p \tilde{i}_p + \frac{n_p}{n_s}\tilde{v}_s - \frac{n_p}{n_s}Z_s \tilde{i}_s \tag{2-13}$$

$$\tilde{v}_s = \frac{n_p}{n_s}\tilde{v}_p - \frac{n_p}{n_s}Z_p \tilde{i}_p + Z_s \tilde{i}_s \tag{2-14}$$

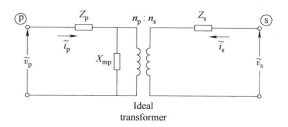

Fig. 2.2 Basic equivalent circuit of a two-winding transformer

$$Z_p = R_p + jX_p; \quad Z_s = R_s + jX_s$$

R_p, R_s = primary and secondary winding resistances

X_p, X_s = primary and secondary winding leakage reactances

n_p, n_s = number of turns of primary and secondary winding

X_{mp} = magnetizing reactance referred to the primary side

Let

$Z_{p0} = Z_p$ at nominal primary side tap position

$Z_{s0} = Z_s$ at nominal secondary side tap position

n_{p0} = primary side nominal number of turns

n_{s0} = secondary side nominal number of turns

Expressing Equations (2-11) and (2-12) in terms of the above nominal values,

$$\tilde{v}_p = \left(\frac{n_p}{n_{p0}}\right)^2 Z_{p0} \tilde{i}_p + \frac{n_p}{n_{p0}} \tilde{v}_s - \left(\frac{n_s}{n_{s0}}\right)^2 Z_{s0} \tilde{i}_s \tag{2-15}$$

$$\tilde{v}_s = \frac{n_s}{n_p} \tilde{v}_p - \frac{n_s}{n_p}\left(\frac{n_p}{n_{p0}}\right)^2 Z_{p0} \tilde{i}_p + \left(\frac{n_s}{n_{s0}}\right)^2 Z_{s0} \tilde{i}_s \tag{2-16}$$

Here, we have assumed that both leakage reactance and resistance of a transformer winding are proportional to the square of the number of turns. This assumption is generally valid for the leakage reactance, but not for the resistance. Since the resistance is much smaller than the leakage reactance and since the deviation of the actual turns ratio from the nominal turns ratio is not very large, the resulting approximation is acceptable. For convenience, we will assume that both primary and secondary windings are connected so as to form a Y-Y connected three-phase bank.

With the nominal number of turns related to the base voltages as follows:

$$\frac{n_{p0}}{n_{s0}} = \frac{v_{pbase}}{v_{sbase}}$$

and

$$v_{pbase} = Z_{pbase} i_{pbase}, \quad v_{sbase} = Z_{sbase} i_{sbase}$$

Equations (2-15) and (2-16) in per unit form become

$$\tilde{v}_p = \tilde{n}_p^2 \tilde{Z}_{p0} \tilde{i}_p + \frac{\tilde{n}_p}{\tilde{n}_s} \tilde{v}_s - \tilde{n}_s^2 \frac{\tilde{n}_p}{\tilde{n}_s} \tilde{Z}_{s0} \tilde{i}_s \tag{2-17}$$

$$\tilde{v}_s = \frac{\tilde{n}_s}{\tilde{n}_p} \tilde{v}_p - \tilde{n}_p^2 \frac{\tilde{n}_s}{\tilde{n}_p} \tilde{Z}_{p0} \tilde{i}_p + \tilde{n}_s^2 \tilde{Z}_{s0} \tilde{i}_s \tag{2-18}$$

where the superbars denote per unit values, with $\tilde{v}_p$, $\tilde{v}_s$, $\tilde{i}_p$, $\tilde{i}_s$ equal to per unit values of phasor voltages and currents, and

$$\tilde{n}_p = \frac{n_p}{n_{p0}} \quad (2\text{-}19)$$

$$\tilde{n}_s = \frac{n_s}{n_{s0}} \quad (2\text{-}20)$$

The per unit equivalent circuit representing Equations (2-17) and (2-18) is shown in Fig. 2.3.

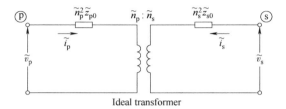

Fig. 2.3 Per unit equivalent circuit

3. Standard Equivalent Circuit

The equivalent circuit of Fig. 2.3 can be reduced to the standard form shown in Fig. 2.4, where n is the per unit turns ratio and it is expressed as

$$\tilde{n} = \frac{\tilde{n}_p}{\tilde{n}_s} = \frac{n_p n_{s0}}{n_{p0} n_s} \quad (2\text{-}21)$$

and

$$\tilde{Z}_e = \tilde{n}_s^2 (\tilde{Z}_{p0} + \tilde{Z}_{s0}) = \left(\frac{n_s}{n_{s0}}\right)^2 (\tilde{Z}_{p0} + \tilde{Z}_{s0}) \quad (2\text{-}22)$$

The equivalent circuit of Fig. 2.4 is widely used for representation of two-winding transformers in power flow and stability studies. The IEEE common format for exchange of solved power flow cases this representation.

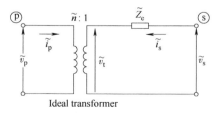

Fig. 2.4 Standard equivalent circuit for a transformer

We see from Equation (2-22) that $\tilde{Z}_e$ does not change with $\tilde{n}_p$. Therefore, if the tap is on the primary side, only $\tilde{n}$ changes.

If the actual turns ratio is equal to n_{p0}/n_{s0}, then $\tilde{n} = 1.0$, and the ideal transformer vanishes.

When the actual turns ratio is not equal to the nominal turns ratio, $\tilde{n}$ represents the off-nominal ratio (ONR).

The equivalent circuit of Fig. 2.4 can be used to represent a transformer with a fixed (or off-load) tap on one side and an ULTC on the other side. The off-nominal turns ratio is assigned to the side with ULTC and Z_e has a value corresponding to the fixed-tap position of the other side, as given by Equation (2-20).

4. Consideration of Three-Phase Transformer Connections

The standard equivalent circuit of Fig. 2.4 represents the single-phase equivalent of a three-phase transformer. In establishing the ONR, the nominal turns ratio (n_{p0}/n_{s0}) is taken to be equal to the ratio of line-to-line base voltages on both sides of the transformer irrespective of the winding connections (Y-Y, D-D, or Y-D).

For Y-Y and D-D connected transformers, this makes the ratios of the base voltages equal to the ratios of the nominal turns of the primary and secondary windings of each transformer phase. For a Y-D connected transformer, this in addition accounts for the factor $\sqrt{3}$ due to the winding connection.

In the case of a Y-D connected transformer, a 30° phase shift is introduced between line-to-line voltages on the two sides of the transformer. The line-to-neutral voltages and line currents are similarly shifted in phase due to the winding connections. It is usually not necessary to take this phase shift into consideration in system studies. Thus, the single-phase equivalent circuit of a Y-D transformer does not account for the phase shift, except in so far as the phase shift of voltages due to the impedance of the transformer.

2.3.3 Representation of Three-Winding Transformers

Fig. 2.5 shows the single-phase equivalent of a three-winding transformer under balanced conditions. The effect of the magnetizing reactance has been neglected, and the transformer is represented by three impedances connected to form a star. The common star point is fictitious and unrelated to the system neutral.

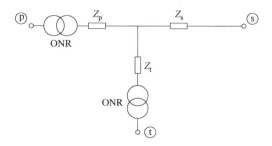

Fig. 2.5 Equivalent circuit of a three-winding transformer

The three windings of the transformer may have different MVA ratings. However, the per unit impedances must be expressed on the same MVA base. As in the case of the two-winding transformer equivalent circuit developed in the previous section, off-nominal turns ratios are used to account for the differences between the ratios of actual turns and the base voltages. The values of the equivalent impedances Z_p, Z_s and Z_t may be obtained by standard short-circuit tests as follows:

Z_{ps} = leakage impedance measured in primary with secondary shorted and tertiary open

Z_{pt} = leakage impedance measured in primary with tertiary shorted and secondary open

Z_{st} = leakage impedance measured in secondary with tertiary shorted and primary open

With the above impedances in ohms referred to the same voltage base, we have

$$Z_{ps} = Z_p + Z_s$$
$$Z_{pt} = Z_p + Z_t$$
$$Z_{st} = Z_t + Z_s \quad (2\text{-}23)$$

Hence

$$Z_p = \frac{1}{2}(Z_{ps} + Z_{pt} - Z_{st})$$

$$Z_s = \frac{1}{2}(Z_{ps} + Z_{st} - Z_{pt})$$

$$Z_t = \frac{1}{2}(Z_{pt} + Z_{st} - Z_{ps})$$

In large transformers, Z_s is small and may even be negative.

2.3.4 Phase-Shifting Transformers

A phase-shifting transformer can be represented by the equivalent circuit shown in Fig. 2.6. It consists of an admittance in series with an ideal transformer having a complex turns ratio, $\tilde{n} = n \angle \alpha$. The phase angle step size may not be equal at different tap positions. However, equal step size is usually used in power flow and transient stability programs.

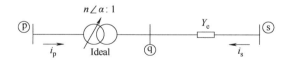

Fig. 2.6 Equivalent circuit of a phase-shifting transformer

By definition:

$$\frac{\tilde{V}_p}{\tilde{V}_q} = n \angle \alpha = n(\cos\alpha + j\sin\alpha) = a_s + jb_s \quad (2\text{-}24)$$

Where α is the phase shift from bus p to bus q; it is positive when $\tilde{v}_p$ leads $\tilde{v}_q$. Since there is no power loss in an ideal transformer,

$$\tilde{v}\tilde{i}_p^* = -\tilde{v}\tilde{i}_s^* \quad (2\text{-}25)$$

Therefore, the transformer current at bus p is

$$\tilde{i}_p = -\frac{1}{a_s - jb_s}\tilde{i}_s = \frac{Y_e}{a_s - jb_s}(\tilde{v}_q - \tilde{v}_s) \quad (2\text{-}26)$$

Substituting for $\tilde{v}_q$ from Equation (2-26), we get

$$\tilde{i}_p = \frac{Y_e}{a_s - jb_s}\left(\frac{1}{a_s + jb_s}\tilde{v}_p - \tilde{v}_s\right) = \frac{Y_e}{a_s^2 + b_s^2}[\tilde{v}_p - (a_s + jb_s)\tilde{v}_s] \tag{2-27}$$

From Equation (2-26),

$$\tilde{i}_s = -(a_s - jb_s)\tilde{i}_p$$

Substituting for $\tilde{i}_p$ from Equation (2-27) gives

$$\tilde{i}_s = \frac{Y_e}{a_s + jb_s}[(a_s + jb_s)\tilde{v}_s - \tilde{v}_p] \tag{2-28}$$

Combining Equations (2-25) and (2-26), we obtain the following matrix equation relating the phase-shifter terminal voltages and currents

$$\begin{pmatrix}\tilde{i}_p \\ \tilde{i}_s\end{pmatrix} = \begin{pmatrix}\dfrac{Y_e}{a_s^2 + b_s^2} & \dfrac{-Y_e}{a_s - jb_s} \\ \dfrac{-Y_e}{a_s + jb_s} & Y_e\end{pmatrix}\begin{pmatrix}\tilde{v}_p \\ \tilde{v}_s\end{pmatrix} \tag{2-29}$$

Note that the admittance matrix in the above equation is not symmetrical, that is, the transfer admittance from p to s is not equal to the transfer admittance from s to p.

If the turns ratio is real (i.e., $a_s = n$ and $b_s = 0$), the model reduces to the equivalent Π circuit shown in Fig. 2.7.

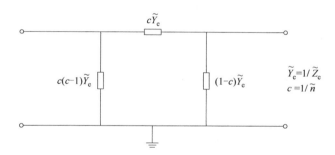

Fig. 2.7 Transformer representation with ONR

2.4 Extra High Voltage (EHV) Transmission

2.4.1 Introduction

At the end of 2005 the installed generation capacity of 500 GW makes China the second highest in the electricity production in the world. 14 years later, in 2019, China generated 7503.4 TW · h of electricity compared to the US's 4401.3 TW · h. China increased the annual generating capacity which was almost twice as much as the US and five times as much as India in third place. As China is being transformed into the manufacturing centre of the world, and the increase of electricity demand is tremendous. New electric energy infrastructure is required to transfer bulk electricity energy over a very long distance. Although the 500 kV AC and 500 kV DC transmission systems are existing, such transmission networks are still considered to be inadequate for bulk electricity energy transmission.

The continuous advances in electric power transmission technology made it possible to transfer bulk electric energy over longer and longer distances. With the development of large scale transmission systems, voltage levels have been increased continuously to achieve greater economies of scale. The highest voltage level now in practical operation is 765 kV. The next voltage level is 1000 kV or over. Basically high voltage (HV) transmission levels include 100 (110) kV, 138 kV, 161 kV, 230 (220) kV, extra high voltage levels (EHV) include 345 (330) kV, 400 kV, 500 kV and 765 (750) kV, and ultra high voltage (UHV) levels are AC voltages higher than 765 kV and DC voltages higher than 600 kV.

The super high voltage grid being considered in China is the transmission grid with voltages higher than 750 kV. It is mainly suitable for transferring bulk electricity energy over a very long distance.

Power industry in developed countries was developing very quickly between the later 1960s and the early 1970s, It was recognized that a higher voltage grid should be introduced when the capacity of power system is doubled. For this reason, the research on super high voltage transmission grid became more important. Many countries, such as the USA, Russia, Italy, France, Japan and Sweden, began to make their design and planning of developing super high voltage grid at that time. However these projects were not completed because of the political and economic factors.

The research project about super high voltage grid in China has been conducted since 1986. But there are fewer projects on engineering application or on a trial basis.

At present, with the development of national economy, the electricity demand has increased very quickly. Noting the fact that there are very long distances between energy bases and major load centers. Therefore, it is necessary to construct the national grid to transfer power from the energy bases to the major load centers. In order to meet the need of society development, the National Grid Corporation has been responsible for the design and planning of construction of the super high voltage power grid.

2.4.2 EHV Transmission in China

In 2004, the electricity shortage occurred in 24 provinces in China, among which Zhejiang, Jiangsu and Fujian as were the most serious ones. The shortage in 2004 was estimated at around 35 GW. One of the reasons is that there were blind investments in power plants while there is a lack of investments in transmission grid.

A series of energy policies for solving the energy problems in the development of coal and electricity in China have been issued by the Chinese government.

China has a vast territory. It is known that there are more resources in hydro and coal, but less in oil and gas. Hydro resources and coal reserves are the most important energy resources in China. However the distribution of electricity energy sources and that of electricity demand are seriously unbalanced.

Nearly two thirds of hydro resources are distributed in the southwest and west of China, including Sichuan, Yunnan and Xizang. Two thirds of the coal reserves are distributed in the northwest and north of China, including Shanxi, Xinjiang and Mongol. On the other hand, two thirds of electricity loads are mainly in the east areas of China where there is a lack of electricity energy sources. The distance between the areas of energy resources and energy demand is very long, say 500-1500 km.

Considering the reasons mentioned above, a higher requirement for the development of an electricity grid is brought forward. According to the present planning, about 120 million kW could be sent to north China, east China and central China from the coal bases in the northwest and north China. The distances between them are about 500-1500 km, which is even more reasonable transportation distance for AC super high voltage grid than that for transportation of energy resources by train since the research results have shown that it is much more economic to transfer the electricity than to transport coal within 1500 km.

It is estimated that the installed generation capability will be 1000 GW in 2020. It has been recognized that large amounts of hydro and coal resources are distributed in the southwest and northwest of China. Hence in order to meet the energy consumption more generating plants will be constructed in these regions by 2020. On the other hand, load centers are situated in the east costal areas of China. According to the experience of international power industry, when the capacity of power grid is expanded by 4 times, a higher voltage level should be introduced. The current 500 kV power grid has a history of 23 years while the installed generation capability in 2004 is around 6 times that of the installed generation capability in 1982. Presently the transmission of electricity across provincial power networks is mainly based on the 500 kV AC and ± 500 kV DC transmission grids. It has been found that such transmission power grids cannot meet the bulk power transmission requirements in 2020. Clearly in this situation power grids of higher voltage levels are required.

Long distance bulk power transmission can satisfy future power growth, secure energy supply, and optimize energy resources allocation. Furthermore, overall social benefits will be improved by enhancing power grid safety and reliability, saving right-of-way, coordinating the development of power plants and grid, alleviating coal transportation pressure and boosting the harmonious development of regional economy.

1000 kV AC and ± 800 kV DC transmission grids have different features. They are complementary. The natural single circuit capacity of 1000 kV AC transmission line is about 5 GW, and that of ±800 kV DC transmission line is at most 6.4 GW. The transmission capacity of DC transmission line depends on the power system connection, restricted by maximum stability limit. Its capacity should be uplifted gradually. In contrast, the transmission capacity of DC transmission line could achieve the design power capacity level as soon as supply source is available.

The reasonable transmission distance of 1000 kV AC transmission line is 1000-2000 km. On the other hand, ± 800 kV DC transmission is the most economic way of long distance power transmission when transmission distance is larger than 1500 km.

1000 kV AC transmission grid, which has general features of AC transmission grid, can form a super backbone grid of a national power system. ± 800 kV DC transmission, "direct super high way", which is complementary to AC transmission grid, can be used to transmission bulk power from large power generating bases to large load centres over long distances. However, ± 800 kV DC transmission will not be suitable for forming a national backbone high voltage power grid due to the limitation of interconnection.

2.4.3 Future of EHV Transmission Grid in China

Based on the technical innovation and upgrading of the current grid, as being planned, the State Grid Corporation is likely to construct the national super high voltage grid, realizing the coordinated development of the grid separated into different voltage levels, optimizing the utilization of energy sources within larger areas, providing safety, economic and high quality electric power services, meeting the requirements of economy and society development.

According to the strategic planning of State Grid Co., the national super high voltage grid, which covers the majority areas of China, should be operated and dispatched through a united national control centre. The national grid is mainly based upon AC interconnected networks with the assistance of DC interconnected networks. The backbone grid will be interconnected by 1000 kV AC transmission with the assistance of ± 800 kV DC transmission. The northwest power grid will be interconnected by 750 kV AC transmission. The first 750 kV AC transmission line with a total length of 140 km was commissioned in September, 2005.

According to the design and planning of the national super high voltage grid, two large capacity transmission channels should be constructed. One is the north-south channel from north via middle to east, forming a huge UHV AC synchronizing power grid. The other is the west-east channel from Sichuan via middle to east. The transmission of the electric energy in northwest coal bases should merge into this super UHV grid channel.

The 1000 kV UHV AC transmission grid should be constructed above 500 kV EHV AC transmission grid, and will be used to transfer the electricity from the coal bases in the northwest and north China to centre and east China, ± 800 kV DC should be used primarily to transfer the electric energy from southwest hydro bases to east areas. Through the national super high voltage backbone grid, coal transportation could be substituted by power transmission. In addition, the united national grid will promote bulk nationwide electricity trading between different regions.

The first UHV project connects North China and Central China, UHV grid will expand to East China. A strong super UHV network will cover North China, Central China and East China. The UHV grid will cover major energy centers and load centers in China. It is estimated that the UHV capacity and trans-regional transfer capacity will exceed 200 GW.

After having constructed 330 kV and 500 kV regional backbone grids, China began to research and evaluate the voltage levels higher than 330 kV and 500 kV in 1980s. Based on the results, it was determined that the next voltage level of 500 kV for AC transmission is 1000 kV, and the next voltage level of 330 kV for AC transmission is 750 kV. The basic research projects about 1000 kV super high voltage transmission grid have been conducted since 1980s. The 750 kV EHV pilot demonstration transmission line was constructed, which would provide operation experience for 1000 kV UHV AC transmission. In addition, the technologies for 1000 kV UHV super high voltage grid have been researched and tested.

As one of the important energy bases, Northwest China plays an important role in the strategy of whole national electricity grid development in order to transfer more power from the western and central China to coastal areas. Northwest electric grid covers Shanxi, Gansu, Qinghai, and Ningxia provinces. There are 1400 km in width from east to west and 900 km in length from south to north. Because of the geographic conditions, the Northwest electric grid covers a larger area, the transmission line is longer, and the distance between energy producing areas and consumption centers is longer. The current 330 kV grid in the northwest grid cannot meet the power exchange requirements. A higher voltage level is needed. After a long-term evaluation, it was shown that the construction of the 750 kV backbones is necessary, feasible and reasonable.

In order to ensure the construction of 750 kV transmission line smoothly, the research project, "the research of key technology for Northwest 750 kV transmission line", was proposed by the State Power Co. (the former of State Grid Co.) in 2001. The major aspects of the project are as follows:

1) The highest operation voltage and the choice of relative technical standards.
2) The research of reactive power compensation of 750 kV system.
3) The research on control methods for 750 kV system load flow and stability.
4) The research of raising the transmission capability of 750 kV line.
5) The research of 750 kV protection relays.

Through the research project, 750 kV grid design, and operation plans were determined.

The first 750 kV pilot demonstration project was commissioned on 26 September, 2005. The project consists of a 750 kV transmission line from Guantin in Qinghai Province to Lanzhou in Gansu Province and two 750 kV substations. The length of the transmission line is 140 km. Now the project has been successfully and safely operated for over 100 days. After the commissioning, the maximum and minimum operating voltages are 777.58 kV and 751 kV, respectively and the maximum power transfer is 316 MW. Up to 4 Jan, 2006, total 254 GW · h electric energy had been transferred. The second 750 kV pilot demonstration project will be launched next year.

In addition to the 750 kV pilot demonstration project, R&D projects for AC 1000 kV UHV pilot project has been carried out heavily:

1) 14 supporting specific researches regarding UHV substation and transmission lines have been carried out.
2) Nearly 50 key technical projects on UHV have been conducted by the end of 2004.
3) 8 key technologies have been reviewed by experts.
4) UHV EM environment indices have been defined.

Feasibility study of the 1000 kV UHV AC pilot project has been carried out. Optimal option for the UHV AC pilot project has been identified. The project now enters preliminary design stage. It is reported that a 1000 kV UHV AC pilot demonstration project, from the south of Anhui Province, via the north of Zhejiang Province, to Shanghai, will be likely to be built in 2008. The total length of the pilot demonstration 1000 kV UHV AC line is 618 km. Along with the 1000 kV UHV AC line, in this project 41000 kV UHV substations will be built.

2.4.4 EHV Grid Effects on the Electricity Markets

Power grid is the base and carrier of electric market. The construction of the national grid is a great project. The design target of the national grid is to fulfill the requirements of power transmission from west to east and from north to south, removing the transmission bottle necks to economic developed areas, optimizing energy source distribution within larger areas, meeting the requirements of economic development, creating competitive nationwide energy markets.

The development of the national electric market should be in pace of that of the national grid. The unified national grid is separated into different layers and regions, which can operate under the same regulations and rules. The electric markets should adapt to the structure of the national grid. According to the structure of the national power grid, the structure of the electricity markets would be separated into three layers, which are national market, regional market and provincial market. The super high voltage grid would make the national electricity market possible and facilitate the nationwide electricity market competition.

VOCABULARY

1. reactive power   无功功率
2. overhead line   架空线
3. leakage current   泄漏电流
4. alternating voltages   交流电压
5. capacitance   n. 电容，电流容量
6. parameter   n. 参数，系数，参量
7. insulate   vt. 隔离，使绝缘
8. impregnate   adj. 浸渍的，浸染的
9. sinusoidal   adj. 正弦曲线的
10. shunt admittance   并联导纳
11. propagation constant   传播常数
12. attenuation constant   衰减常数
13. incident voltage   入射电压
14. exponential   adj. 指数的
15. tap   n. 抽头
16. compensate   vt. 补偿
17. tap-changing   n. 抽头切换
18. distribution transformers   输电变压器
19. autotransformer   n. 自耦变压器
20. magnetic coupling   磁力耦合
21. transformation ratio   电压比
22. overload   n. 过载，超负荷
23. saturation   n. 饱和，磁化饱和
24. turns ratio   匝数比
25. deviation   n. 偏差，误差
26. irrespective of   不考虑
27. fictitious   adj. 虚构的，假想的
28. transient   adj. 暂态的
29. symmetrical   adj. 匀称的，对称的
30. transfer admittance   转移导纳
31. tremendous   adj. 巨大的
32. infrastructure   n. 基础设施
33. transmission grid   输电网
34. generating plants   发电厂

35. optimize　　*vt.* 使最优化
36. reliability　*n.* 可靠性
37. coordinate　*vi.* 协调
38. alleviate　*vt.* 减轻，缓和
39. complementary　*adj.* 补足的，补偿的
40. innovation　*n.* 创新
41. commissioned　*adj.* 受委任的，被任命的
42. synchronize　*vt.* 使同步
43. backbone grids　主干网
44. demonstration　*n.* 示范，证明
45. geographic　*adj.* 地理的，地理学的
46. substation　*n.* 变电站
47. feasibility　*n.* 可行性
48. preliminary　*adj.* 初步的，预备的
49. facilitate　*vt.* 促进，帮助

NOTES

1. For overhead lines, the inductances of the three phases are different from each other unless the conductors have equilateral spacing, a geometry not usually adopted in practice.

对于架空线路，除非导体间距相等，否则三相间电感互不相同，但在实际中通常不采用这种（等边三角形）结构。

2. Underground cables have the same basic parameters as overhead lines: series resistance and inductance; shunt capacitance and conductance.

地下电缆具有与架空线路相同的基本参数：串联电阻和电感；并联电容和电导。

3. In modern electric power systems, the transmitted power undergoes four to five voltage transformations between the generators and the ultimate consumers. Consequently, the total MVA rating of all the transformers in a power system is about five times the total MVA rating of all the generators.

在现代电力系统中，发电机和最终用户之间的传输功率经历四到五级电压变换。因此，在电力系统中所有变压器的总供电量额定值大约是所有发电机总供电量的五倍。

4. In addition to voltage transformation, transformers are often used for control of voltage and reactive power flow. Therefore, practically all transformers used for bulk power transmission and many distribution transformers have taps in one or more windings for changing the turns ratio.

除了电压变换，变压器经常用于控制电压和无功潮流。因此，几乎所有用于大功率传输的变压器和许多配电变压器都有一个或多个分接头来改变匝数比。

5. As compared to the conventional two-winding transformer, the autotransformer has advantages of lower cost, higher efficiency, and better regulation. These advantages become less significant as the transformation ratio increases; hence, autotransformers are used for low transformation ratios (for example, 500/230 kV).

相比传统的双绕组变压器，自耦变压器具有成本低、效率高、调节好等优点。这些优势在高电压比的时候并不是很明显，因此，自耦变压器用于低电压比（例如，500/230 kV）。

6. As China is being transformed into the manufacturing center of the world, and the increase of electricity demand is tremendous. New electric energy infrastructure is required to transfer bulk electricity energy over a very long distance. Although the 500 kV AC and 500 kV DC transmission systems are existing, such transmission networks are still considered to be inadequate for bulk

electricity energy transmission.

中国正在变成世界制造业中心,并且电力需求迅速增加。需要远距离大容量电能传输新的电网架构。尽管有500 kV交流和500 kV直流的输电系统,但该网络仍然被认为不能满足大容量的电能传输。

7. It is known that there are more resources in hydro and coal, but less in oil and gas. Hydro resources and coal reserves are the most important energy resources in China. However the distribution of electricity energy sources and that of electricity demand are seriously unbalanced.

众所周知,在中国水资源和煤炭资源比石油和天然气资源丰富。水资源和煤炭储量是中国最重要的能源资源。然而电力能源和电力需求的分布严重失衡。

8. The reasonable transmission distance of 1000 kV AC transmission line is 1000-2000 km. On the other hand, ±800 kV DC transmission is the most economic way of long distance power transmission when transmission distance is larger than 1500 km.

1000 kV交流输电线路合理的传输距离为1000～2000 km。另一方面,当传输距离超过1500 km时,±800 kV直流输电系统是长距离输电最经济的方式。

9. According to the strategic planning of State Grid Co., the national super high voltage grid, which covers the majority areas of China, should be operated and dispatched through a united national control centre. The national grid is mainly based upon AC interconnected networks with the assistance of DC interconnected networks.

根据国家电网公司战略规划,覆盖中国大部分地区的超高压电网会通过一个统一的国家控制中心进行操作和调度。国家电网依托交流互联为主、直流互联为辅进行互联。

10. According to the structure of the national power grid, the structure of the electricity markets would be separated into three layers, which are national market, regional market and provincial market. The super high voltage grid would make the national electricity market possible and facilitate the nationwide electricity market competition.

根据国家电网的结构,电力市场的结构分为国家、区域和省三层。超高压电网将使全国电力市场成为可能,并促进全国范围内的电力市场竞争。

Chapter 3
Distribution Network

3.1 Introduction

Electricity distribution is the final stage in the delivery of electricity to end users. A distribution system network carries electricity from the transmission system and delivers to consumers. Typically, the network would include medium-voltage (less than 50 kV) power lines, substation and pole-mounted transformers, low-voltage (less than 1 kV) distribution wiring and sometimes electricity meters.

Conductors for distribution may be carried on overhead pole lines, or in densely-populated areas where they are buried underground. Urban and suburban distribution is done with three-phase systems to serve both residential, commercial, and industrial loads. Distribution in rural areas may be only single-phase if it is not economic to install three-phase power for relatively few and small customers.

Only large consumers are fed directly from distribution voltages. Most utility customers are connected to a transformer, which reduces the distribution voltage to the relatively low voltage used by lighting and interior wiring systems. The transformer may be pole-mounted or set on the ground in a protective enclosure. In rural areas, a pole-mount transformer may serve only one customer, but in more built-up areas multiple customers may be connected. In very dense city areas, a secondary network may be formed with many transformers feeding into a common bus at the utilization voltage. Each customer has an "electrical service" or "service drop" connection and a meter for billing. (Some very small loads, such as yard lights, may be too small to meter and so are charged only a monthly rate.)

A ground connection to local earth is normally provided for the customer's system as well as for the equipment owned by the utility. The purpose of connecting the customer's system to ground is to limit the voltage that may develop if high voltage conductors fall on the lower-voltage conductors, or if a failure occurs within a distribution transformer. If all conductive objects are bonded to the same earth grounding system, the risk of electric shock is minimized. However, multiple connections between the utility ground and customer ground can lead to stray voltage problems; customer piping, swimming pools or other equipment may develop objectionable voltages. These problems may be difficult to resolve since they often originate from places other than the customer's premises.

In many areas, "delta" three-phase service is common. Delta service has no distributed neutral wire and therefore it is less expensive. In North America and Latin America, three-phase service is often a Y (wye) in which the neutral is directly connected to the center of the generator rotor. The neutral provides a low-resistance metallic return to the distribution transformer. Wye service is recognizable when a line has four conductors, one of which is lightly insulated. Three-phase wye service is excellent for motors and heavy power use.

Many areas in the world use single-phase 220 V or 230 V residential and light industrial service. In this system, the high voltage distribution network supplies a few substations per area, and the 230 V power from each substation is directly distributed. A hot wire and neutral are connected to the building from one phase of three-phase service. Single-phase distribution is used where motor loads are small.

In the U. S., parts of Canada and Latin America, split phase service is the most common. Split phase provides both 120 V and 240 V service with only three wires. The house voltages are provided by local transformers. The neutral is directly connected to the three-phase neutral. Socket voltages are only 120 V, but 240 V is available for heavy appliances because the two halves of a phase oppose each other.

Japan has a large number of small industrial manufacturers, and therefore supplies standard low-voltage three-phase service in many suburbs. Also, Japan normally supplies residential service as two phases of a three-phase service, with a neutral. They work well for both lighting and motors.

Rural services normally try to minimize the number of poles and wires. Single-wire earth return (SWER) is the least expensive, with one wire. It uses high voltages, which in turn permit use of galvanized steel wire. The strong steel wire permits inexpensive wide pole spacings. Other areas use high voltage split-phase or three-phase service at higher cost.

Electricity meters use different metering equations depending on the form of electrical service. Since the math differs from service to service, the number of conductors and sensors in the meters also vary.

Besides referring to the physical wiring, the term "electrical service" also refers in an abstract sense to the provision of electricity to a building.

In this chapter, distribution network structure, load model and distribution automation system will be introduced.

3.2 Distribution Systems Structure

3.2.1 Primary Distribution Systems

The wiring between the generating station and the final distribution point is called the primary distribution system. There are several methods used for transmitting the power between these two points. The two most common methods are the radial system and the loop system.

1. The Radial System

The term radial comes from the word radiate, which means to send out or emit from one central point. A radial system is an electrical transmission system which begins at a central station and supplies power to various substations.

In its simplest form, a radial system consists of a generating station which produces the electrical energy. This energy is transmitted from the generator (s) to the central station, which is generally part of, or adjacent to, the generating station. At the central station, the voltage is stepped up to a higher value for long-distance transmission.

From the central station, several lines carry the power to various substations. At the substations, the voltage is usually lowered to a value more suitable for distribution to populated areas. From the substations, lines carry the power to distribution transformers. These transformers lower the voltages to the value required by the consumer.

2. The Loop System

The loop system starts from the central station or a substation and makes a complete loop through the area to be served, and back to the starting point. This results in the area being supplied from both ends, allowing sections to be isolated in case of a breakdown. An expanded version of the loop system consists of several central stations joined together to form a very large loop.

3.2.2 Consumer Distribution Systems

The type of distribution system that the consumer uses to transmit power within the premises depends upon the requirements of the particular installation. Residential occupancies generally use the simplest type. Commercial and industrial systems vary widely with load requirements.

1. Single-phase Systems

Most single-phase systems are supplied from a three-phase primary. The primary of a single-phase transformer is connected to one phase of the three-phase system. The secondary contains two coils connected in series with a midpoint tap to provide a single-phase, three-wire system. This arrangement is generally used to supply power to residential occupancies and some commercial establishments. A schematic diagram is shown in Fig. 3.1.

For residential occupancies, the service conductors are installed either overhead or underground. Single-family and small multifamily dwellings have the kilowatt-hour meters installed on the outside of the building. From the kilowatt-hour meter, the conductors are connected to the main disconnect. Figs. 3.2a and b show this arrangement.

Three separate disconnecting means are used with one common ground.

From the main disconnect, the conductors supply power to the branch circuit panels. For dwelling occupancies, there are three basic types of branch circuits: general lighting circuits, small appliance and laundry circuits, and individual branch circuits. The individual branch circuits are frequently used to supply central heating and/or air-conditioning systems, water heaters, and other special loads.

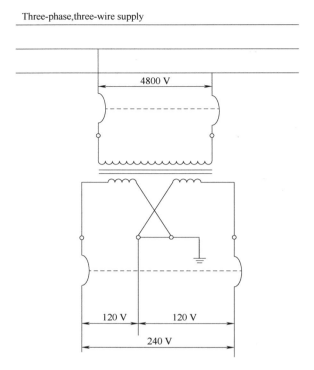

Fig. 3.1 Single-phase, three-wire, 120/240 V system

2. Grounding Requirements

All AC services are required to be grounded on the supply side of the service disconnecting means. This grounding conductor runs from the combination system and equipment ground to the grounding electrode. For multifamily occupancies, it is permitted to use up to six service disconnecting means. A single grounding conductor of adequate size should be used for the system ground (see Fig. 3.2b).

3.2.3 Commercial and Industrial Installations

Commercial and industrial installations are more complex than small residential installations. Large apartment complexes and condominiums, although classified as residential occupancies, often use commercial-style services. A single-phase, three-wire service or a three-phase, four-wire service may be brought into the building, generally from underground. The service-entrance conductors terminate is a main disconnect. From this point, the conductors are connected to the individual kilowatt-hour meters for each apartment and then to smaller disconnecting means and over-current protective devices. Branch-circuit panels are generally installed in each apartment. Feeder conductors connect the individual disconnecting means to the branch-circuit panels. Commercial and/or industrial buildings may have more than one kilowatt-hour meter, depending upon the number of occupancies. The service sizes vary according to the demand. The service is usually a three-phase, four-wire system. The available voltages may be 120/208 V or 277/480 V. If the

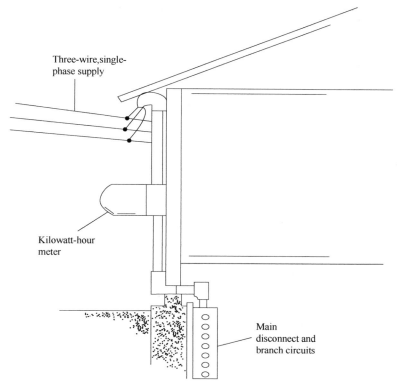

a) Single-family residence with a three-wire, single-phase supply

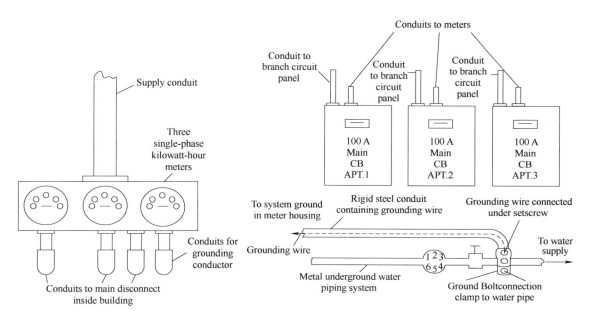

b) Three-wire, single-phase service for a multifamily dwelling

Fig. 3. 2   Kilowatt-hour meter installed for residential occupancies

system provides 277/480 V, a transformer must be installed in order to obtain 120 V. If the building covers a large area, it is recommended that the service must be installed near the center of the building. This arrangement minimizes line loss on feeder and branch-circuit conductors. Some utilities supply a three-phase, three-wire or three-phase, four-wire delta system. The common voltages that may be obtained from the three-wire delta system are 240 V, 440 V, or 550 V. With this arrangement, a transformer must be used to obtain 120 V. The usual voltages supplied from the four-wire delta system are 240 V, three phases and 120 V, single phase.

Many large consumers purchase the electrical energy at the primary voltage, and transformers are installed on their premises. Three-phase voltages up to 15 kV are often used.

The service for this type of installation generally consists of metal cubicles called a substation unit. The transformers are either installed within the cubicle or adjacent to it. Isolation switches of the drawer type are installed within the cubicle. These switches are used to isolate the main switch or circuit breaker from the supply during maintenance or repair.

3.2.4 Consumer Loop System

Although the radial system of distribution is probably the most commonly used system of transmitting power on the consumer's property, the loop system is also employed. A block diagram of the consumer loop system is illustrated in Fig. 3.3. There are several variations of the loop system in use in the industry, but the system illustrated here show the basic structure.

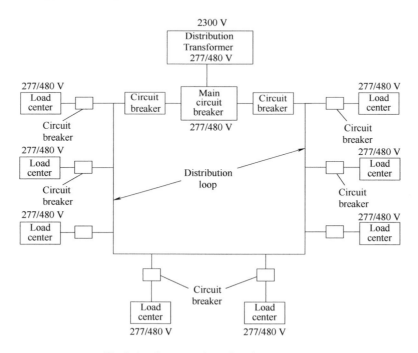

Fig. 3.3 Consumer loop distribution system

When installing any system, over-current protection and grounding must be given primary consideration. Electrical personnel who design and install these systems must comply with the NEC and local requirements. Disconnecting means may be installed anywhere in the distribution loop to provide for isolating sections.

3.2.5 Secondary High-Voltage Distribution

Large industrial establishments may find it more economical to distribute power at voltages higher than 600 V. Depending upon the type of installation and the load requirements, voltages as high as 2300 V may be used. Step-down transformers are installed in strategic locations to reduce the voltage to a practical working value. A diagram of a high-voltage radial system is shown in Fig. 3.4.

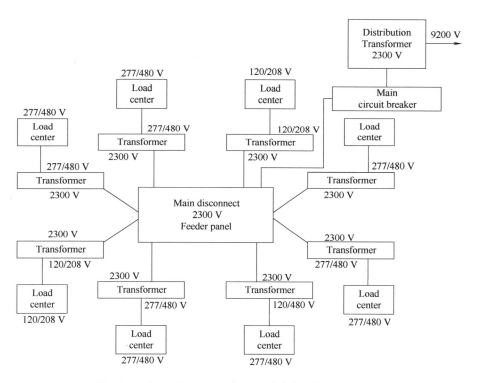

Fig. 3.4 Secondary high-voltage radial distribution system

Sometimes the high-voltage (primary) system may be radial, and the low-voltage (secondary) system may be connected into a loop. Another method is to have both the primaries and secondaries connected to form a loop. Figs. 3.5a and b show these methods.

3.2.6 Secondary Ties Loop System

It is frequently convenient to connect loads to the secondary conductors at points between transformers. These conductors are called secondary ties. Article 450 of the NEC gives specific requirements regarding the conductor sizes and over-current protection.

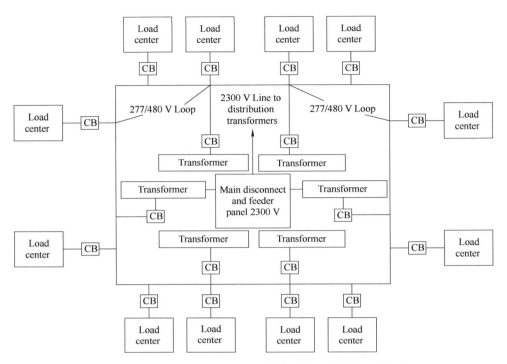

a) Secondary high-voltage distribution system: high-voltage radical; low-voltage loop

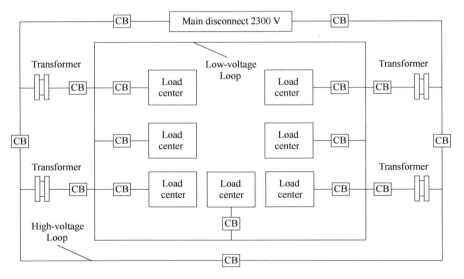

b) Consumer distribution system with high-voltage and low-voltage loops

Fig. 3.5　Different ties of high-voltage and low-voltage

3.3 Load in Distribution System

In power system stability and power flow studies, the common practice is to represent the composite load characteristics as seen from bulk power delivery points. As illustrated in Fig. 3.6, the aggregated load represented at a transmission substation (Bus A) usually includes, in addition to the connected load devices, the effects of substation step-down transformers, subtransmission feeders, distribution feeders, distribution transformers, voltage regulators and reactive power compensation devices.

The load models are traditionally classified into two broad categories: static models and dynamic models.

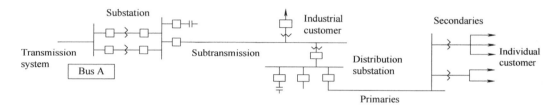

Fig. 3.6 Power system configuration identifying parts of the system represented as load at a bulk power delivery point (Bus A)

3.3.1 Static Load Models

A static load model expresses the characteristics of the load at any instant of time as algebraic functions of the bus voltage magnitude and frequency at that instant. The active power component P and the reactive power component Q are considered separately.

Traditionally, the voltage dependency of load characteristics has been represented by the exponential model:

$$P = P_0 (\overline{V})^a$$
$$Q = Q_0 (\overline{V})^b \tag{3-1}$$

In this and other load models described in this section,

$$\overline{V} = \frac{V}{V_0}$$

Where P and Q are active and reactive components of the load when the bus voltage magnitude is V. The subscript 0 identifies the values of the respective variables at the initial operating condition.

The parameters of this model are the exponents a and b. With these exponents equal to 0, 1, or 2, the model represents constant power, constant current, or constant impedance characteristics, respectively. For composite loads, their values depend on the aggregate characteristics of load components.

The exponent a (or b) is nearly equal to the slope dP/dV (or dQ/dV) at $V = V_0$. For composite system loads, the exponent a usually ranges between 0.5 and 1.8; the exponent b is typically between 1.5 and 6. A significant characteristic of the exponent b is that it varies as a nonlinear function of voltage. This is caused by magnetic saturation in distribution transformers and motors. At higher voltages, Q tends to be significantly higher.

In the absence of specific information, the most commonly accepted static load model is to represent active power as constant current (i.e. $a = 1$) and reactive power as constant impedance (i.e. $b = 2$).

An alternative model which has been widely used to represent the voltage dependency of loads is the polynomial model:

$$\begin{cases} P = P_0(p_1 \overline{V}^2 + p_2 \overline{V} + p_3) \\ Q = Q_0(q_1 \overline{V}^2 + q_2 \overline{V} + q_3) \end{cases} \quad (3\text{-}2)$$

This model is commonly referred to as the ZIP model, as it is composed of constant impedance (Z), constant current (I), and constant power (P) components. The parameters of the model are the coefficients p_1 to p_3 and q_1 to q_3, which define the proportion of each component.

The frequency dependency of load characteristics is usually represented by multiplying the exponential model or the polynomial model by a factor as follows:

$$\begin{cases} P = P_0(\overline{V})^a (1 + K_{pf}\Delta f) \\ Q = Q_0(\overline{V})^b (1 + K_{qf}\Delta f) \end{cases} \quad (3\text{-}3)$$

or

$$\begin{cases} P = P_0(p_1 \overline{V}^2 + p_2 \overline{V} + p_3)(1 + K_{pf}\Delta f) \\ Q = Q_0(q_1 \overline{V}^2 + q_2 \overline{V} + q_3)(1 + K_{qf}\Delta f) \end{cases} \quad (3\text{-}4)$$

Where Δf is the frequency deviation ($f - f_0$). Typically, K_{pf} ranges from 0 to 3.0, and K_{qf} ranges from -2.0 to 0. The bus frequency f is usually not a state variable in the system model used for stability analysis. Therefore, it is evaluated by computing the time derivative of the bus voltage angle.

A comprehensive static model which offers the flexibility of accommodating several forms of load representation is as follows:

$$P = P_0(P_{ZIP} + P_{EX1} + P_{EX2}) \quad (3\text{-}5)$$

where

$$\begin{cases} P_{ZIP} = p_1 \overline{V}^2 + p_2 \overline{V} + p_3 \\ P_{EX1} = p_4 (\overline{V})^{a1} (1 + K_{pf1}\Delta f) \\ P_{EX2} = p_5 (\overline{V})^{a2} (1 + K_{pf2}\Delta f) \end{cases} \quad (3\text{-}6)$$

The expression for the reactive component of the load has a similar structure. The reactive power compensation associated with the load is represented separately.

The static models given by Equations (3-1) to (3-6) are not realistic at low voltages, and may lead to computational problems. Therefore, stability programs usually make provisions for switching the load characteristic to the constant impedance model when the bus voltage falls below a specified value. In the load model used in the EPRI extended transient/midterm stability program (ETMSP), the exponents a_1, a_2, b_1, and b_2 are varied as a function of voltage below a threshold value of bus voltage, and the constant power and constant current components are switched to constant impedance representation.

3.3.2 Dynamic Load Models

The response of most composite loads to voltage and frequency changes is fast, and the steady state of the response is reached very quickly. This is true at least for modest amplitudes of voltage/frequency change. The use of static models described in the previous sections is justified in such cases.

There are, however, many cases where it is necessary to account for the dynamics of load components. Studies of interarea oscillations, voltage stability, and long-term stability often require load dynamics to be modeled. Study of systems with large concentrations of motors also requires representation of load dynamics.

Typically, motors consume 60% to 70% of the total energy supplied by a power system. Therefore, the dynamics attributable to motors are usually the most significant aspects of dynamic characteristics of system loads.

Other dynamic aspects of load components that require consideration in stability studies include the following:

1) Extinction of discharge lamps below a certain voltage and their restart when the voltage recovers. Discharge lamps include mercury vapour, sodium vapour, and fluorescent lamps. These extinguish at voltages in the range of 0.7 p.u. to 0.8 p.u. When the voltage recovers, they restart after 1 s or 2 s delay.

2) Operation of protective relays, such as thermal and over-current relays. Many industrial motors have starters with electromagnetically held contactors. These drops open at voltages in the range of 0.55 p.u. to 0.75 p.u.; the dropout time is on the order of a few cycles. Small motors on refrigerators and air conditioners have only thermal overload protections, which typically trip in about 10 s to 30 s.

3) Thermostatic control of loads, such as space heaters/coolers, water heaters, and refrigerators. Such loads operate longer during low-voltage conditions. As a result, the total number of these devices connected to the system will increase in a few minutes after a drop in voltage. Air conditioners and refrigerators also exhibit such characteristics under sustained low-frequency conditions.

4) Response of ULTCs on distribution transformers, voltage regulators, and voltage-controlled capacitor banks. These devices are not explicitly modeled in many studies. In such cases, their effects must be implicitly included in the equivalent load that is represented at the bulk power delivery point.

As these devices restore distribution voltages following a disturbance, the power supplied to voltage sensitive loads is restored to the pre-disturbance levels. The control action begins about 1 minute after the change in voltage, and the voltage restoration within the capability of these devices is completed in a total time of 2 min to 3 min.

A composite load model which allows the representation of the wide range of characteristics exhibited by the various load components is shown in Fig. 3.7. The model has provision for representing aggregations of small induction motors, large induction motors, static load characteristics described in Equations (3-5) and (3-6), discharge lighting, thermostatically controlled loads, transformer saturation effects, and shunt capacitors.

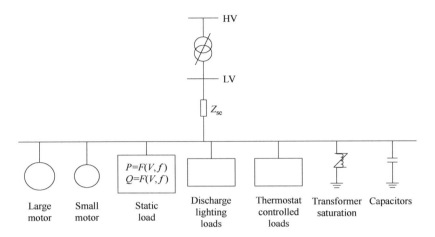

Fig. 3.7 Composite static and dynamic load model

1. Thermostatically Controlled Load

Fig. 3.8 shows a simple model of the aggregated thermostatically controlled (that is, constant energy) loads. In this model, G is the load conductance, G_0 is the initial value of G, $K_L G_0$ is the maximum value of G which represents a condition with all such loads on, and T is a time constant.

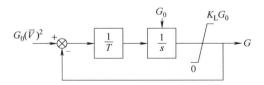

Fig. 3.8 A simple model for thermostatically controlled loads

A more realistic model of the thermostatically controlled loads used in the ETMSP is shown in Fig. 3.9. The basis for the model is as follows:

The dynamic equation of a heating device may be written as

$$K\frac{\mathrm{d}\tau_H}{\mathrm{d}t} = P_H - P_L \tag{3-7}$$

Where

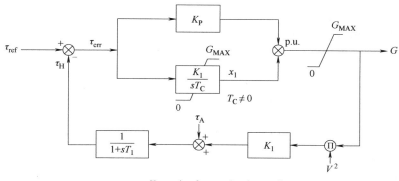

K_P = gain of proportional controller
K_f = gain of integral controller
T_C = time constant of integral controller,s
τ_{ref} = reference temperature
τ_A = ambient temperature
T_1 = load time constant,s
K_1 = gain associated with load model
G_0 = initial value of G
G_{MAX} = maximum value of G

Fig. 3. 9 A realistic model for themostatically controlled loads

τ_H = temperature of heated area

τ_A = ambient temperature

P_H = power from the heater

$\quad = K_H G V^2$

P_L = heat loss by escape to ambient area

$\quad = K_A (\tau_H - \tau_A)$

G = load conductance

Substituting the expressions for P_H and P_L in Equation (3-7) gives

$$K \frac{d\tau_H}{dt} = K_H G V^2 - K_A (\tau_H - \tau_A) \qquad (3\text{-}8)$$

Rearranging, we have

$$\frac{d\tau_H}{dt} = \frac{K_H}{K} G V^2 + \frac{K_A}{K} \tau_A - \frac{K_A}{K} \tau_H$$

or

$$\frac{d\tau_H}{dt} = \frac{K_1}{T_1} G V^2 + \frac{1}{T_1} \tau_A - \frac{1}{T_1} \tau_H \qquad (3\text{-}9)$$

Where

$$T_1 = \frac{K}{K_A}$$

and

$$K_1 = \frac{K_H}{K_A}$$

The temperature τ_H is compared with the reference temperature, and the error signal controls the load conductance through a proportional plus integral controller. When all the thermostatically controlled loads supplied by the load bus are on, G reaches its maximum value G_{MAX}.

From Fig. 3.9, under pre-disturbance conditions τ_H is equal to τ_{ref}. Hence,

$$\tau_{ref} = K_1 V_0^2 G_0 + \tau_A$$

or

$$K_1 = \frac{\tau_{ref} - \tau_A}{V_0^2 G_0} \tag{3-10}$$

2. Discharge Lighting Load

Fig. 3.10 shows a model suitable for representing the characteristics of discharge lighting loads in stability studies. At bus voltages less than V_1, the lamps extinguish. For voltages greater than V_1, P and Q vary as nonlinear functions of V.

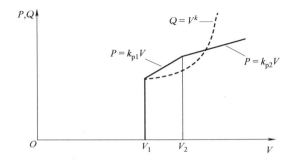

Fig. 3.10 Discharge lighting characteristics

3.4 Distribution Network Automation System

3.4.1 Introduction of the Distribution Network Automation System

The characteristics and typical scheme designs of distribution automation system (DAS) of northeast rural county cities of China are presented after intercommunication with many electric power corporations and making designs for them. The DASs are intended to increase the reliability of power supply after reforms of the distribution network. The "local control type" with upgrade capability to "central control type" is chosen in benefit of inexpensiveness and simpleness. The function requirements include fault determination, isolation and restoration of healthy area for hand-by-hand lines and radial lines. About the characteristics, electric switches of voltage-time type are mainly adopted; the uniform specification of installment and setting of reclosers is provided, such as installing a recloser at the original end of each feeder line and resetting existing reclosers; typical modes of tie switches are presented; the simple communication methods and high performance devices are adopted for the severe winters of northeast areas.

The reliability of power supply becomes more demanding with the gradual development of economics, and reforms in the distribution network in many northeast rural areas have been conducted accordingly, such as cable reform, transformer substation automatization and etc.. Then many counties commence constructing distribution network automation systems based on these reforms. Considering the lower demand of power reliability in rural counties than in main cities as well as the limit bankroll, distribution automation system of local control type is implemented in general instead of central control type. This system should perform the required functions of fault determination, isolation and healthy area restoration, and accommodate hand-by-hand feeder lines and radial feeder lines.

According to the characteristics of rural distribution network, automated switches of voltage-time type are adopted in general, and some have the capability of over-current protection or over-current locking. Some reclosers may be already installed in the middle or the terminal of original lines, and the setting value or working mode of those reclosers should be adjusted under new scheme. Furthermore, some lines include several reclosers under new scheme and their over-current and time setting values should be set with care to make them coordinate in one feeder line.

The tie switch has several working modes as following: the tie switches connect common hand-by-hand lines; two tie switches lie in three feeder lines, the tie switches between two transformer substations coexist with the tie switches of feeder lines, and the tie switches in trunk feeder lines coexist with the tie switches in sub-feeder lines. Theoretically, automatic switch action could be realized through setting different closing time delay, i. e. XL-TIME. But allowing for the length of the maximum closing time delay and logic complexity, some tie switches such as those between two transformer substations are generally operated manually.

To fulfill the demands under the severe winter in northeast rural areas, FTUs (Feeder Terminal Unit) and batteries should be selected with care, which should function properly till $-50℃$. Switch FTU may as well have some simple communication functions, in order that dispatchers can monitor line faults.

3.4.2 Control Type Selection for Distribution Network Automation

Current distribution network automation system involves two control types: central control type and local control type. The system of central control type integrates field switch and FTU data into master station through a communication system, and performs centralized analysis, control and optimization. Whereas the system of local control type, without master station and corresponding communication system, automatically performs the required functions of fault determination, isolation and healthy area restoration mostly through preset coordination among switches.

Though central control type is the trend of distribution network automation system, its extensive application in rural areas is unpractical at this stage. The difficulties lie in:

1) Most counties have difficulty with the investment in fiber communication system and master system because of their limited bankroll. Surely, some counties with ample funds hope to construct central control type systems.

2) Management level does not keep step with the level of distribution network automation

system. Many staffs have no clear comprehension on distribution network automation, and the amount of technicians is insufficient in distribution network maintenance, automation and computer. Most counties have not established departments specialized in distribution network automation.

3) The installation and adjustment of central control type system is very time-consuming and complex considering the severe winters in northeast areas, most counties lack experience in these aspects.

4) Power networks in counties are simply structured in general, then many advanced functions in central control type system are of redundancy such as power flow analysis and reactive power optimization.

5) Reclosers or circuit breakers with over-current protection have been installed at feeder lines in many counties, then it is of great waste to replace them.

Distribution network automation system of local control type is appropriate for many counties, which requires no communication system and master station system. The primary components in this system are switches including reclosers and load break switches. The operations of these switches can be performed automatically and no manual field operations are needed. Basic functions of distribution network automation can be realized in this system. Furthermore many power corporations in counties require upgrade capability to central control type system through selecting appropriate switches.

3.4.3 Switch Type and Control Method in Distribution Network

In the distribution network automation system of central control type, SF_6 switches of electric type and FTUs with protection and communication functions are adopted in genera. Therefore electric SF_6 switches with intelligent controllers are adopted in order to be upgraded conveniently. These intelligent controllers are used to conduct some control schemes of local control type. They have some simplified functions of measuring and logic determination and can be regarded as simplified FTUs, but they do not support protocol communication and remote control functions. Switches have three operating modes: radial feeder line mode, hand-by-hand normally close mode, and hand-by-hand normally open mode (i.e. tie switch).

To realize the scheme of local control type conveniently, reduce logic complexity and improve reliability, the operating mode of switches at trunk feeder lines are of voltage-time type, i.e. "close when power supplied and trip when out-of-voltage". With respect to the switches at the tail end of radial lines or sub-feeder lines, they may support the operating mode of current-time type. But over-current locking should be set for them, i.e. close locking is activated when a switch is in over-current state before tripping upon out-of-voltage, and closing when power supplied is not permitted to prevent incorrect locking of switches at trunk feeder lines. With respect to the switches at hand-by-hand trunk feeder lines, they can modify operating mode between normally close mode and normally open mode automatically, and can adjust all protective setting values such as closing time delay automatically when power supply direction is reversed.

Recloser is important in distribution network automation system of local control type. Some reclosers with over-current protection have been already installed at some feeder lines in northeast areas, and exporting switches of feeder lines are reclosers in most transformer substations after reforms for automatization. To improve the utilization efficiency of existing equipments and uniform the configuration of feeder lines, the typical scheme involves two situations: ①install a recloser at the original end (outside transformer substation) of a feeder line; ②append at most one recloser at a trunk feeder line, so at most two reclosers are installed at a trunk feeder line. Excessive reclosers will make the protective setting complicated.

About the coordination between the exporting recloser of a feeder line inside transformer substation and the recloser at the original end of a feeder line, there are two methods: ①Modify the current and time setting of the recloser inside transformer substation, and make these values greater than the values of the recloser at the original end of a feeder line; ②remove relaying protection function of the recloser inside transformer substation, and make it into a common circuit breaker. These methods are helpful for assignment of responsibility in distribution network. Equipments inside transformer substation are managed by dispatching department, while feeder line equipments outside transformer substation are managed by distribution network department. Fig. 3.11 shows two reclosers and one sectionalizer with over-current locking function at a radial feeder line.

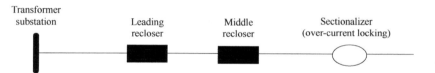

Fig. 3.11 Typical configurature of recloses

3.4.4 Operation of Tie Switch in Distribution Network

Hand-by-hand feeder line is the primary form in distribution network automation. Lacking of bankroll, many county power corporations in northeast areas expect every switch to play its maximum value and maximize capability of power restoration after faults. Therefore the designs of tie switches have some difference compared with those in main city distribution network. There are four primary forms: ①one tie switches between two feeder lines, the most common form; ②two tie switches in three feeder lines; ③the tie switches between two transformer substations coexisting with the tie switches in feeder lines; ④the tie switches in trunk feeder lines coexisting with the tie switches in sub-feeder lines. These forms are illustrated in Fig. 3.12 and Fig. 3.13.

In Fig. 3.12, if only the part "two tie switches in three feeder lines" is considered, power restoration may be performed through line 1 or line 3 when line 2 is in fault, so the reliability is improved. The closing time delays of these two tie switches should be set in difference of 5 s-10 s. Once one tie switch closes and power is restored successfully, the other tie switch does not close. An alternative scheme is adding a tie switch between line 1 and line 3 and improves the reliability of three lines further, but it is unpractical because of its logic complexity. In like manner, with respect to the tie switch between two transformer substations in Fig. 3.12, the automation of switches can be realized by setting different closing time delays in theory. But allowing for the length of the maximum closing time delay and logic complexity, the tie switches between two transformer substations are generally operated manually.

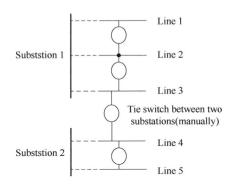

Fig. 3.12 Tie switches between two transformer substations; two switches in three lines

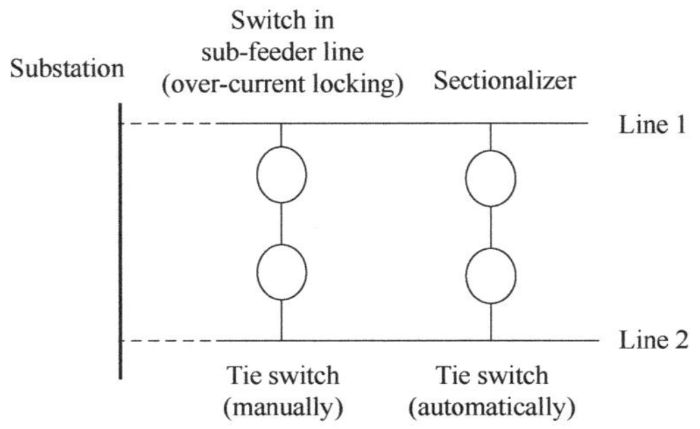

Fig. 3.13 Tie switches in trunk feeder lines and sub-feeder lines

On occasion, in order to change the operation mode of some feeder lines conveniently or guarantee power supply for easily faulted sections, the tie switches in trunk feeder lines may coexist with the tie switches in sub-feeder lines. Considering that concurrent closing of two tie switches is likely to induce a loop, which is not permitted, one of them is generally set to operate manually to avoid complex logic.

3.4.5 Components Selections in DAS

The temperature in the severe winter in northeast areas may reach -50℃, and the scatterance of power distribution equipments in rural areas is inconvenient to dispose of faults in time. Therefore switches and companioned equipments are required to be of high performance. The components of industrial or military grade in intelligent controllers should be selected to avoid malfunctions under low temperature.

Batteries, companioned with a switch, are important component to successful fault processing, which supply power for a FTU and drive the corresponding switch. But batteries are sensitive to low temperature, so high performance batteries should be selected to guarantee no breakage, driving force to switches and long life time. Cyclon serial batteries of USA Enersys Company are adopted after comparison with batteries of other corporations, which is kind of valve-regulated lead acid battery with absorbing glass mat.

To improve capability of batteries further, a heat preservation equipment is employed. It warms batteries when a feeder supplies the power. With respect to the cold start under low temperature, if a feeder is powered down in a long perod, batteries will stop power supplying to avoid overdischarge, the FTU will stop working and its temperature is equal to environment temperature, thus FTU may not conduct a successful cold start once the feeder is powered again. Therefore heating equipment is employed to warm FTU to an appropriate temperature at feeder power restoration, and then FTU is started and batteries return to work.

3.4.6 Simplified Supervising System in DAS

Feeder lines in some counties lie in rural areas, far away from county power corporations. Timely feeder checking and fault eliminating are difficult to be realized in winter. Therefore power corporations hope to obtain the information about working state and fault location in office. Considering that SMS (Short Messaging Service) communication has advantages on large covering range, technique maturity, communication reliability and low cost, a SMS module is added to a FTU, and a SMS processing system is set at master station which polls every FTU through SMS. Though the communication speed and reliability of SMS communication is lower than those of fiber communication, this is a simplified scheme and is convenient to be realized. By this means, dispatchers could monitor the analog and digital values of a switch, and determine the fault section through monitoring the switches tripped by fault. Then technicians can conduct appropriate field maintenance and working efficiency is promoted.

VOCABULARY

1. residential   *adj.* 住宅的，与住宅有关的
2. interior   *adj.* 内部的，室内的
3. enclosure   *n.* 外壳
4. meter   *n.* 仪表；*vt.* 用表测量
5. conductive   *adj.* 传导的，导电的
6. generator rotor   发电机转子
7. split phase   分相
8. socket   *n.* 插座
9. galvanized steel wire   镀锌钢丝
10. generating station   发电厂
11. adjacent to   邻近的，毗连的
12. loop system   闭环系统
13. disconnect   *n.* 断开处
14. grounding electrode   接地电极
15. over-current protective devices   过电流保护装置
16. feeder   *n.* 馈线，支线
17. radial system   放射状系统
18. comply with   遵守
19. step-down transformer   降压变压器
20. load characteristic   负载特性，负载特性曲线
21. distribution feeder   配电馈线
22. voltage regulator   电压调节器
23. active power   有功功率
24. reactive power   无功功率
25. aggregate   *adj.* 聚合的，集合的
26. magnetic saturation   磁性饱和
27. polynomial   *adj.* 多项式的
28. frequency deviation   频率偏差
29. stability analysis   稳定性分析
30. threshold   *n.* 阈值，临界值，极限
31. amplitude   *n.* 振幅
32. oscillation   *n.* 振动，摆动
33. attributable to   由于，可归咎于
34. fluorescent lamps   荧光灯
35. protective relay   保护继电器
36. contactor   *n.* 开关，接触器
37. equivalent load   等效负荷
38. reference temperature   参考温度，基准温度
39. lighting load   电光负载，照明负载
40. recloser   *n.* 重合闸
41. setting value   设定值

42. dispatcher *n.* 调度员，分配器
43. master station 主站，主控台，总机
44. scatterance 散布
45. analog *adj.* 模拟的
46. module *n.* 模块

NOTES

1. Only large consumers are fed directly from distribution voltages. Most utility customers are connected to a transformer, which reduces the distribution voltage to the relatively low voltage used by lighting and interior wiring systems.

只有大容量用户直接连接到配电网。大多数用户连接到变压器上，变压器把配电网的电压降到用于照明和室内布线的相对低等级的电压。

2. This energy is transmitted from the generator (s) to the central station, which is generally part of, or adjacent to, the generating station. At the central station the voltage is stepped up to a higher value for long-distance transmission.

电能从发电机传输到中心电站，中心电站通常是发电厂的一部分或在发电厂附近。中心电站将电压升到更高的等级用于远距离传输。

3. The service-entrance conductors terminate is a main disconnect. From this point, the conductors are connected to the individual kilowatt-hour meters for each apartment and then to smaller disconnecting means and over-current protective devices.

进线口导线端子是一个主要的断开点。从这一点开始，连接到每户公寓的出线都经过各自的电度表，然后再接小容量的断开装置和过电流保护装置。

4. Depending upon the type of installation and the load requirements, voltages as high as 2300 V may be used. Step-down transformers are installed in strategic locations to reduce the voltage to a practical working value.

根据安装类型和负荷需求，可能需要 2300 V 的电压。在关键位置安装降压变压器将电压降低到实际工作电压值。

5. In power system stability and power flow studies, the common practice is to represent the composite load characteristics as seen from bulk power delivery points.

在电力系统稳定性和潮流研究中，一般的做法是用总的功率传输来描述复合负荷特性。

6. In the load model used in the EPRI extended transient/midterm stability program (ETMSP), the exponents a_1, a_2, b_1, and b_2 are varied as a function of voltage below a threshold value of bus voltage, and the constant power and constant current components are switched to constant impedance representation.

在 EPRI 扩展暂态/中长期稳定程序中的负荷模型中，当电压方程中的电压低于母线电压阈值时，通过改变变量 a_1、a_2、b_1、b_2 的值，恒功率和恒电流元件可以转为恒阻抗（模型）。

7. The reliability of power supply becomes more demanding with the gradual development of economics, and reforms in the distribution network in many northeast rural areas have been conducted accordingly, such as: cable reform, transformer substation automatization and etc..

随着经济的逐步发展，对供电可靠性的要求变得越来越苛刻，一些东北偏远农村地区的

配电网也进行了相应的升级改造，比如线路电缆化和变电站自动化等。

8. According to the characteristics of rural distribution network, automated switches of voltage-time type are adopted in general, and some have the capability of over-current protection or over-current locking. Some reclosers may be already installed in the middle or the terminal of original lines, and the setting value or working mode of those reclosers should be adjusted under new scheme.

根据农村配电网络的特点，电压-时间型自动开关被普遍采用，一些会有过电流保护或过电流锁定能力。那些早已安装在原来线路中间或者末端的自动重合闸，其设定值或工作模式在新方案下应进行（相应的）调整。

9. Whereas the system of local control type, without master station and corresponding communication system, automatically performs the required functions of fault determination, isolation and healthy area restoration mostly through preset coordination among switches.

然而局部控制型系统在没有主站和相应通信系统的情况下，通过开关之间的预置协调，自动执行故障判定、隔离和恢复稳定区域的要求功能。

10. Distribution network automation system of local control type is appropriate for many counties, which requires no communication system and master station system. The primary components in this system are switches including reclosers and load break switches. The operations of these switches can be performed automatically and no manual field operations are needed.

局部控制型配电网自动化系统适合多个地区，这些地区不需要通信系统和主站系统。该系统的主要部分是开关，包括重合闸和断路器。这些开关可以进行自动操作而且无须人工干预。

11. With respect to the switches at hand-by-hand trunk feeder lines, they can modify operating mode between normally close mode and normally open mode automatically, and can adjust all protective setting values such as closing time delay automatically when power supply direction is reversed.

对于"手拉手"主干馈线开关，可以在常闭模式和常开模式之间自动修改其操作模式，而且在电源方向相反的情况下可以自动调整所有的保护设定值，如闭合延时。

12. On occasion, in order to change the operation mode of some feeder lines conveniently or guarantee power supply for easily faulted sections, the tie switches in trunk feeder lines may coexist with the tie switches in sub-feeder lines.

某些情况下，为了更方便地改变馈线操作模式或保证故障区域的供电，主馈线的联络开关可以和分支馈线的联络开关共同存在。

13. With respect to the cold start under low temperature, if a feeder is powered down in a long period, batteries will stop power supplying to avoid overdischarge, the FTU will stop working and its temperature is equal to environment temperature, thus FTU may not conduct a successful cold start once the feeder is powered again.

至于低温情况下的冷启动，如果馈线长时间失电，电池将停止供电以避免过度放电，FTU将停止工作，其温度将等于周围环境温度。这样的话，当馈线再一次通电，FTU可能无法成功进行冷启动。

Chapter 4
Measurement and Control in Power System

4.1 Introduction

Measurement and control instruments have played a very important role in making the power system in a normal and safe operation. There are several kinds of measurement instrument such as electricity meter, phase measurement unit (PMU), remote terminal unit (RTU), protective relay, measurement and control units etc.. Most of the measurement and control instruments achieve their functions by the data acquisition (DAQ) technology with data signal processor (DSP).

Traditional measurement and protection systems consist of current and voltage transformers (CTs and PTs) that produce analog signals that are transported through electrical cables to the substation control room. The primary sides of the instrument transformers are supplied by the network voltage and current. The secondary side supplies the protection relays or energy measurement systems. The conventional instrument transformers have one output, 1 A or 5 A for CTs, and 69 V or 115 V for PTs.

The hardware on each measurement and control station was developed around a signal conditioning and instrumentation system to allow the system to perform tasks. The measurement hardware on each instrument consists the secondary side of voltage and current sensors, signal-conditioning boards (SCBs), and a DAQ system. The secondary side voltage attenuators transform three-phase and neutral voltages from up to 250 V to levels acceptable to the DAQ system (< 10 V) at 100∶1 ratio. These voltage attenuators have a typical gain error of +0.035%, a maximum gain error of +0.08%, and an offset error of +0.51 mV/gain respect to input (RTI) (with respect to the input). The secondary side three-phase and neutral currents are measured using 100 A Liaisons Electroniques-Mécaniques (LEM) LA-100P hall-effect current transducers, which have a 2000∶1 ratio and a 0.45% accuracy. The attenuated voltage and current signals are then sent to multichannel SCBs. The signal-conditioning circuit has four stages and provides the following functions: attenuating signals, suppressing voltage spikes, preventing ground loops, and filtering high-frequency noise over 2000 Hz. A schematic of the signal-conditioning circuit is shown in Fig. 4.1.

Repetitive testing was performed to establish the SCBs' output-to-input ratios. The average ratio for the four voltage channels was found to be 0.985, while the ratio for the four current channels

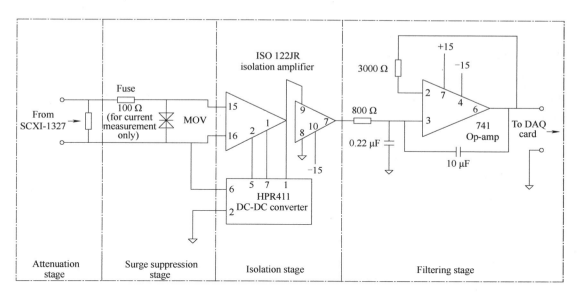

Fig. 4.1 Voltage and current signal-conditioning circuits

was determined to be 1.002. Each measurement and control instrument is programmed to capture up to 96 filtered voltage and current signals using a 16-b 500 kHz DAQ system.

Recently developed optical voltage and current transducers produce digital and analog signals representing the secondary current and voltage. These signals are transported to the control room through fiber-optic cables. In the control room, the optical instrument transformer may supply digital relays, a digital energy metering system, power quality meters, among other devices or systems. The optical instrument transformers have three outputs: digital; low energy analog (LEA): optical CT = 4 V, optical PT = 4 V; and high energy analog (HEA): optical CT = 1 or 5 A, optical PT = 69/120 V. The advantages of optical systems over traditional magnetic systems include improved safety, smaller size, immunity from electromagnetic interferences, better transient response (wider frequency band), a broader dynamic range, and higher accuracy. It is expected that the optical and digital system will provide advantages in relaying and metering applications. The wider frequency band and improved dynamic performance should make the relaying applications less prone to misoperation as a result of transformer saturation or input signal distortions. The higher accuracy should result in more reliable revenue metering. The wider frequency bandwidth should also improve accuracy of harmonic monitoring, which would enable improved assessment of power quality. The immunity from electromagnetic disturbances should increase reliability.

Supervisory control and data acquisition (SCADA) systems have been used in the utilities industry in the United States since the 1960s. These systems are used to monitor critical infrastructure systems and provide early warning of potential disaster situations. In the electric utility industry, SCADA usually refers to basic control and monitoring of field devices including breakers, switches, capacitors, reclosers, and transformers.

In the 1950s, analog communications were employed to collect real-time data of MW power outputs from power plants and tie-line flows to power companies for operators using analog computers to conduct load frequency control (LFC) and economic dispatch (ED). Using system frequency as a surrogate measurement of power balance between generation and load within a control area, LFC was used to control generation in order to maintain frequency and interchange schedules between control areas. An ED adjusts power outputs of generators at equal incremental cost to achieve overall optimality of minimum total cost of the system to meet the load demand. Penalty factors were introduced to compensate for transmission losses by the loss formula. This was the precursor of the modern control center. When digital computers were introduced in the 1960s, remote terminal units (RTUs) were developed to collect real-time measurements of voltage, real and reactive powers, and status of circuit breakers at transmission substations through dedicated transmission channels to a central computer equipped with the capability to perform necessary calculation for automatic generation control (AGC), which is a combination of LFC and ED. Command signals to remotely raise or lower generation levels and open or close circuit breakers could be issued from the control center. This is called the SCADA system.

After the northeast blackout of 1965, a recommendation was made to apply digital computers more extensively and effectively to improve the real-time operations of the interconnected power systems. The capability of control centers was pushed to a new level in the 1970s with the introduction of the concept of system security, covering both generation and transmission systems. The security of a power system is defined as the ability of the system to withstand disturbances or contingencies, such as generator or transmission line outages. Because security is commonly used in the sense of against intrusion, the term power system reliability is often used today in place of the traditional power system security in order to avoid causing confusion to laymen. The security control system is responsible for monitoring, analysis, and real-time coordination of the generation and the transmission systems. It starts from processing the telemetered real-time measurements from SCADA through a state estimator to clean out errors in measurements and communications. Then the output of the state estimator goes through the contingency analysis to answer "what-if" questions. Contingencies are disturbances such as generator failure or transmission line outages that might occur in the system. This is carried out using a steady-state model of the power system, i.e., power flow calculations. Efficient solution algorithms for large nonlinear programming problem known as the optimal power flow (OPF) were developed for transmission-constrained economic dispatch, preventive control, and security-constrained ED (SCED). Due to daily and weekly variations in load demands, it is necessary to schedule the startup and shutdown of generators to ensure that there is always adequate generating capacity on-line at minimum total costs. The optimization routine doing such scheduling is called unit commitment (UC). Control centers equipped with state estimation and other network analysis software, called Advanced Application Software, in addition to the generation control software, are called energy management systems (EMS). Early control centers used specialized computers offered by vendors whose business was mainly in the utility industry. Later, general purpose computers, from mainframe to mini, were used to do SCADA, AGC, and security control. In the late 1980s minicomputers were gradually replaced by a set of UNIX workstations or PCs running on an LAN. At the same time, SCADA systems were installed in substations and distribution feeders. More functions were added step by step to these distribution management systems (DMS) as the computational power of PCs increased.

4.2 SCADA System in Power System

4.2.1 SCADA Overview

SCADA is an acronym for supervisory control and data acquisition. SCADA systems are used to monitor and control a plant or equipment in industries such as telecommunications, water and waste control, energy, oil and gas refining and transportation. These systems encompass the transfer of data between a SCADA central host computer and a number of remote terminal units (RTUs) and/or programmable logic controllers (PLCs), and the central host and the operator terminals. A SCADA system gathers information (such as where a leak on a pipeline has occurred), transfers the information back to a central site, then alerts the home station that a leak has occurred, carries out necessary analysis and control, such as determining if the leak is critical, and displaying the information in a logical and organized fashion. These systems can be relatively simple, such as one that monitors environmental conditions of a small office building, or very complex, such as a system that monitors all the activity in a nuclear power plant or the activity of a municipal water system. Traditionally, SCADA systems have made use of the public switched network (PSN) for monitoring purposes.

Today many systems are monitored using the infrastructure of the corporate local area network (LAN)/wide area network (WAN). Wireless technologies are now being widely deployed for purposes of monitoring.

SCADA systems consist of:

1) One or more field data interface devices, usually RTUs, or PLCs, which interface to field sensing devices and local control switchboxes and valve actuators.

2) A communications system used to transfer data between field data interface devices and control units and the computers in the SCADA central host. The system can be radio, telephone, cable, satellite, etc., or any combination of these.

3) A central host computer server or servers sometimes called a SCADA center, master station, or master terminal unit (MTU).

4) A collection of standard and/or custom software sometimes called human machine interface (HMI) software or man machine interface (MMI) software systems used to provide the SCADA central host and operator terminal application, support the communications system, and monitor and control remotely located field data interface devices.

Fig. 4.2 shows a very basic SCADA system, while Fig. 4.3 shows a typical SCADA system. Each of the above system components will be discussed in detail in the next sections.

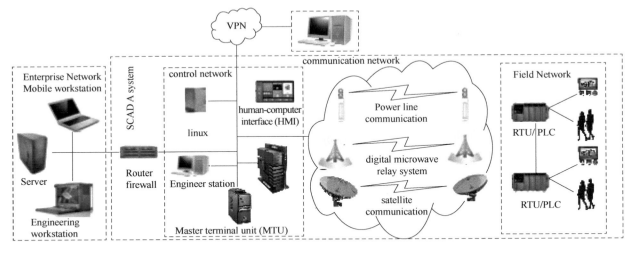

Fig. 4.2　Basic SCADA system

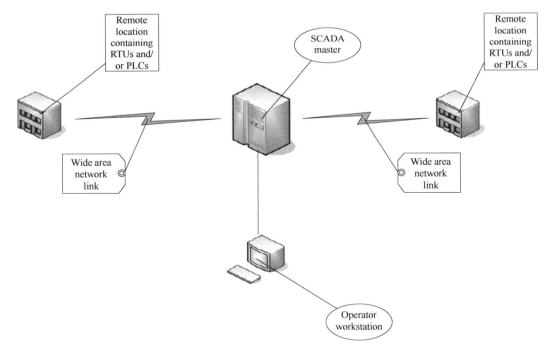

Fig. 4.3　Typical SCADA system

1. Field Data Interface Devices

Field data interface devices form the "eyes and ears" of a SCADA system. Devices such as reservoir level meters, water flow meters, valve position transmitters, temperature transmitters, power consumption meters, and pressure meters all provide information that can tell an experienced operator how well a water distribution system is performing. In addition, equipment such as electric valve actuators, motor control switchboards, and electronic chemical dosing facilities can be used to form the "hands" of the SCADA system and assist in automating the process of distributing water.

However, before any automation or remote monitoring can be achieved, the information that is passed to and from the field data interface devices must be converted to a form that is compatible with the language of the SCADA system. To achieve this, some form of electronic field data interface is required. RTUs, also known as remote telemetry units, provide this interface. They are primarily used to convert electronic signals received from field interface devices into the language (known as the communication protocol) used to transmit the data over a communication channel.

The instructions for the automation of field data interface devices, such as pump control logic, are usually stored locally. This is largely due to the limited bandwidth of communications links between the SCADA central host computer and the field data interface devices. Such instructions are traditionally held within the PLCs, which have been physically separate from RTUs in the past. A PLC is a device used to automate monitoring and control of industrial facilities. It can be used as a stand-alone or in conjunction with a SCADA or other system. PLCs connect directly to field data interface devices and incorporate programmed intelligence in the form of logical procedures that will be executed in the event of certain field conditions.

PLCs have their origins in the automation industry and therefore are often used in manufacturing and process plant applications. The need for PLCs to connect to communication channels was not great in these applications, as they often were only required to replace traditional relay logic systems or pneumatic controllers. SCADA systems, on the other hand, have origins in early telemetry applications, where it was only necessary to know basic information from a remote source. The RTUs connected to these systems had no need for control programming because the local control algorithm was held in the relay switching logic.

As PLCs were used more often to replace relay switching logic control systems, telemetry was used more and more with PLCs at the remote sites. It became desirable to influence the program within the PLC through the use of a remote signal. This is in effect the "supervisory control" part of the acronym SCADA. Where only a simple local control program was required, it became possible to store this program within the RTU and perform the control within that device. At the same time, traditional PLCs included communications modules that would allow PLCs to report the state of the control program to a computer plugged into the PLC or to a remote computer via a telephone line. PLC and RTU manufacturers therefore compete for the same market.

As a result of these developments, the line between PLCs and RTUs has blurred and the terminology is virtually interchangeable. For the sake of simplicity, the term RTU will be used to refer to a remote field data interface device; however, such a device could include automation programming that traditionally would have been classified as a PLC.

2. Communications Network

The communications network is intended to provide the means by which data can be transferred between the central host computer servers and the field-based RTUs. The communication network refers to the equipment needed to transfer data to and from different sites. The medium used can either be cable, telephone or radio. The use of cable is usually implemented in a factory. This is not practical for systems covering large geographical areas because of the high cost of the cables, conduits and the extensive labor in installing them. The use of telephone lines (i.e. leased or dial-up) is a more economical solution for systems with large coverage. The leased line is used for systems requiring on-line connection with the remote stations. This is expensive since one telephone line will be needed per site. Dial-up lines can be used on systems requiring updates at regular intervals (e.g. hourly updates). Here ordinary telephone lines can be used. The host can dial a particular number of a remote site to get the readings and send commands.

Remote sites are usually not accessible by telephone lines. The use of radio offers an economical solution. Radio modems are used to connect the remote sites to the host. An on-line operation can also be implemented on the radio system. For locations wherein a direct radio link cannot be established, a radio repeater is used to link these sites.

Historically, SCADA networks have been dedicated networks; however, with the increased deployment of office LANs and WANs as a solution for interoffice computer networking, there exists the possibility to integrate SCADA and LANs into everyday office computer networks.

The foremost advantage of this arrangement is that there is no need to invest in a separate computer network for SCADA operator terminals. In addition, there is an easy path to integrating SCADA data with existing office applications, such as spreadsheets, work management systems, data history databases, geographic information system (GIS) systems, and water distribution modeling systems.

3. **Central Host Computer**

The central host computer or master station is most often a single computer or a network of computer servers that provide a man-machine operator interface to the SCADA system. The computers process the information received from and sent to the RTU sites and present it to human operators in a form that the operators can work with. Operator terminals are connected to the central host computer by a LAN/WAN so that the viewing screens and associated data can be displayed for the operators. Recent SCADA systems are able to offer high resolution computer graphics to display a graphical user interface or mimic screen of the site or water supply network in question. Historically, SCADA vendors offered proprietary hardware, operating systems, and software that was largely incompatible with other vendors' SCADA systems. Expanding the system required a further contract with the original SCADA vendor. Host computer platforms characteristically employed UNIX-based architecture, and the host computer network was physically removed from any office-computing domain.

However, with the increased use of the personal computer, computer networking has become commonplace in the office and as a result, SCADA systems are now available that can network with office-based personal computers. Indeed, many of today's SCADA systems can reside on computer servers that are identical to those servers and computers used for traditional office applications. This has opened a range of possibilities for the linking of SCADA systems to office-based applications such as GIS systems, hydraulic modeling software, drawing management systems, work scheduling systems, and information databases.

4. Operator Workstations and Software Components

Operator workstations are most often computer terminals that are networked with the SCADA central host computer. The central host computer acts as a server for the SCADA application, and the operator terminals are clients that request and send information to the central host computer based on the request and action of the operators.

An important aspect of every SCADA system is the computer software used within the system. The most obvious software component is the operator interface or man machine interface/human machine interface (MMI/HMI) package; however, software of some form pervades all levels of a SCADA system. Depending on the size and nature of the SCADA application, software can be a significant cost item when developing, maintaining, and expanding a SCADA system. When software is well defined, designed, written, checked, and tested, a successful SCADA system will likely be produced. Poor performances in any of these project phases will very easily cause a SCADA project to fail.

Many SCADA systems employ commercial proprietary software upon which the SCADA system is developed. The proprietary software often is configured for a specific hardware platform and may not interface with the software or hardware produced by competing vendors. A wide range of commercial off-the-shelf (COTS) software products also are available, some of which may suit the required application. COTS software usually is more flexible, and will interface with different types of hardware and software.

Generally, the focus of proprietary software is on processes and control functionality, while COTS software emphasizes compatibility with a variety of equipment and instrumentation. It is therefore important to ensure that adequate planning is undertaken to select the software systems appropriate for any new SCADA system.

Software products typically used within a SCADA system are as follows:

1) Central host computer operating system: Software used to control the central host computer hardware. The software can be based on UNIX or other popular operating systems.

2) Operator terminal operating system: Software used to control the central host computer hardware. The software is usually the same as the central host computer operating system. This software, along with that for the central host computer, usually contributes to the networking of the central host and the operator terminals.

3) Central host computer application: Software that handles the transmittal and reception of data to and from the RTUs and the central host. The software also provides the graphical user interface which offers site mimic screens, alarm pages, trend pages, and control functions.

4) Operator terminal application: Application that enables users to access information available on the central host computer application. It is usually a subset of the software used on the central host computers.

5) Communications protocol drivers: Software that is usually based within the central host and

the RTUs, and is required to control the translation and interpretation of the data between ends of the communications links in the system. The protocol drivers prepare the data for use either at the field devices or the central host end of the system.

6) Communications network management software: Software required to control the communications network and to allow the communications networks themselves to be monitored for performance and failures.

7) RTU automation software: Software that allows engineering staff to configure and maintain the application housed within the RTUs (or PLCs). Most often this includes the local automation application and any data processing tasks that are performed within the RTU.

The preceding software products provide the building blocks for the application-specific software, which must be defined, designed, written, tested, and deployed for each SCADA system.

4.2.2 SCADA Architectures

SCADA systems have evolved in parallel with the growth and sophistication of modern computing technology. The following sections will provide a description of the following three generations of SCADA systems:

1) First generation: Monolithic;
2) Second generation: Distributed;
3) Third generation: Networked.

1. Monolithic SCADA Systems

When SCADA systems were first developed, the concept of computing in general centered on "mainframe" systems. Networks were generally non-existent, and each centralized system stood alone. As a result, SCADA systems were standalone systems with virtually no connectivity to other systems.

The WANs that were implemented to communicate with RTUs were designed with a single purpose in mind—that of communicating with RTUs in the field and nothing else. In addition, WAN protocols in use today were largely unknown at the time.

The communication protocols in use on SCADA networks were developed by vendors of RTU equipment and were often proprietary. In addition, these protocols were generally very "lean", supporting virtually no functionality beyond that required scanning and controlling points within the remote device. Also, it was generally not feasible to intermingle other types of data traffic with RTU communications on the network.

Connectivity to the SCADA master station itself was very limited by the system vendor. Connections to the master typically were done at the bus level via a proprietary adapter or controller plugged into the central processing unit (CPU) backplane.

Redundancy in these first generation systems was accomplished by the use of two identically equipped mainframe systems, a primary and a backup, connected at the bus level. The standby system's primary function was to monitor the primary and take over in the event of a detected failure. This type of standby operation meant that little or no processing was done on the standby system. Fig. 4.4 shows a typical first generation SCADA architecture.

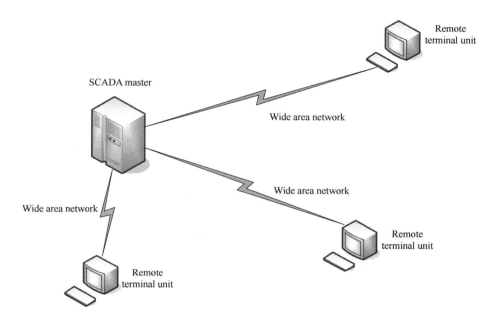

Fig. 4.4 First generation SCADA architecture

2. Distributed SCADA Systems

The next generation of SCADA systems took advantage of developments and improvement in system miniaturization and LAN technology to distribute the processing across multiple systems. Multiple stations, each with a specific function, were connected to a LAN and shared information with each other in real-time. These stations were typically of the mini-computer class, smaller and less expensive than their first generation processors.

Some of these distributed stations served as communications processors, primarily communicating with field devices such as RTUs. Some served as operator interfaces, providing the HMI for system operators. Still others served as calculation processors or database servers. The distribution of individual SCADA system functions across multiple systems provided more processing power for the system as a whole than would have been available in a single processor. The networks that connected these individual systems were generally based on LAN protocols and were not capable of reaching beyond the limits of the local environment.

Some of the LAN protocols that were used were of a proprietary nature, where the vendor created its own network protocol or version thereof rather than pulling an existing one off the shelf. This allowed a vendor to optimize its LAN protocol for real-time traffic, but it limited (or effectively eliminated) the connection of the network from other vendors to the SCADA LAN. Fig. 4.5 depicts typical second generation SCADA architecture.

Distribution of system functionality across network-connected systems served not only to increase

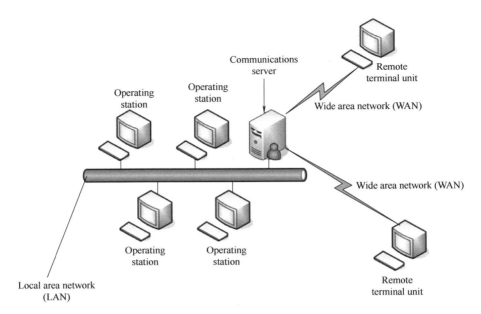

Fig. 4.5 Second generation SCADA architecture

processing power, but also to improve the redundancy and reliability of the system as a whole. Rather than the simple primary/standby fail over scheme that was utilized in many first generation systems, the distributed architecture often kept all stations on the LAN in an online state all of the time. For example, if a HMI station were to fail, another HMI station could be used to operate the system, without waiting for fail over from the primary system to the secondary.

The WAN used to communicate with devices in the field was largely unchanged by the development of LAN connectivity between local stations at the SCADA master. These external communications networks were still limited to RTU protocols and were not available for other types of network traffic.

As was the case with the first generation of systems, the second generation of SCADA systems was also limited to hardware, software, and peripheral devices that were provided or at least selected by the vendor.

3. Networked SCADA Systems

The current generation of SCADA master station architecture is closely related to that of the second generation, with the primary difference being that of an open system architecture rather than a vendor controlled, proprietary environment. There are still multiple networked systems, sharing master station functions. There are still RTUs utilizing protocols that are vendor-proprietary. The major improvement in the third generation is that of opening the system architecture, utilizing open standards and protocols and making it possible to distribute SCADA functionality across a WAN and not just a LAN.

Open standards eliminate a number of the limitations of previous generations of SCADA systems. The utilization of off-the-shelf systems makes it easier for the user to connect third party peripheral devices (such as monitors, printers, disk drives, tape drives, etc.) to the system and/or the network.

As they have moved to "open" or "off-the-shelf" systems, SCADA vendors have gradually gotten out of the hardware development business. These vendors have looked to system vendors such as Compaq, Hewlett-Packard, and Sun Microsystems for their expertise in developing the basic computer platforms and operating system software. This allows SCADA vendors to concentrate their development in an area where they can add specific value to the system—that of SCADA master station software.

The major improvement in third generation SCADA systems comes from the use of WAN protocols such as the internet protocol (IP) for communication between the master station and communications equipment. This allows the portion of the master station that is responsible for communications with the field devices to be separated from the master station "proper" across a WAN. Vendors are now producing RTUs that can communicate with the master station using an Ethernet connection. Fig. 4.6 represents a networked SCADA system.

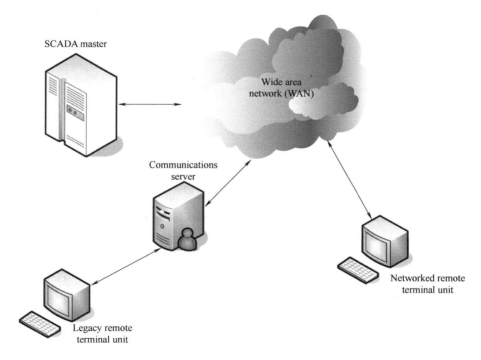

Fig. 4.6　Third generation SCADA system

Each protocol consists of two message sets or pairs. One set forms the master protocol, containing the valid statements for master station initiation or response, and the other set is the RTU protocol, containing the valid statements that an RTU can initiate and respond to. In most but not all cases, these pairs can be considered a poll or request for information or action and a confirming response.

The SCADA protocol between master and RTU forms a viable model for RTU-to-intelligent electronic device (IED) communications. Currently, in industry, there are several different protocols in use. The most popular protocols are international electrotechnical commission (IEC) 60870-5 series, specifically IEC 60870-5-101 (commonly referred to as 101) and distributed network protocol version 3 (DNP3).

IEC 60870-5 specifies a number of frame formats and services that may be provided at different layers. IEC 60870-5 is based on a three-layer enhanced performance architecture (EPA) reference model (see Fig. 4.7) for efficient implementation within RTUs, meters, relays, and other IEDs.

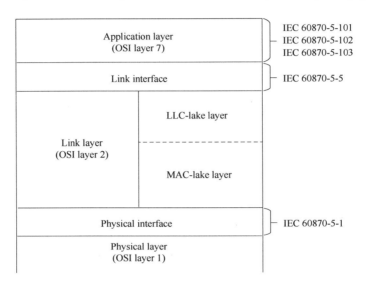

Fig. 4.7 EPA reference model

Additionally, IEC 60870-5 defines basic application functionality for a user layer, which is situated between the open system interconnection (OSI) application layer and the application program. This user layer adds interoperability for such functions as clock synchronization and file transfers. The following descriptions provide the basic scope of each of the five documents in the base IEC 60870-5 telecontrol transmission protocol specification set.

Standard profiles are necessary for uniform application of the IEC 60870-5 standards. A profile is a set of parameters defining the way a device acts. Such profiles have been and are being created. The 101 profile is described in detail following the description of the applicable standards.

4.3 Power System Control

4.3.1 Key Concepts in Protection and Control

Protection is very closely related to circuit-breaker trip signals to disconnect faulty or overloaded equipment from the network, to save the component, and to re-establish normal operation of the healthy part of the power system and thereby continue electricity supply to the customers. Protection equipment is also aimed at protecting people, animals, and property from injury and damage due to electric faults. The situation when a protection device triggers is so severe that if the equipment is not tripped, it will be severely damaged or the surroundings will be exposed to serious danger.

System protection or wide-area protection is used to save the system from a partial or total blackout or brownout in operational situations when no particular equipment is faulted or operated outside its limitations. This situation could appear after the clearance of a very severe disturbance in a stressed operation situation or after an extreme load growth. Since it is a protection system, it will operate in such operational situations when the power system would break down if no protective actions were taken. Such protective actions also comprise shifts of setting groups and parameter values for different protection and control devices, blocking of tap-changers, switch in of shunt capacitors, etc.

Emergency control is associated with continuous control actions in order to save the power system, such as boosting the exciter on a synchronous generator or changing the power direction of a HVDC link.

Normal control actions are associated with continuous control activities, that can be either stepwise, e. g., tap-changer and shunt device, or continuous, such as frequency control. Normal control is preventive, i. e., actions are taken to adjust the power system operational conditions to the present and near future expected situation. Normal control is usually automatic, e. g., tap-changer, reactive shunt device, frequency control, and AGC.

The difference between normal and emergency control is the consequence for the power system if the control action is not performed. If a normal, preventive, control action is not performed, there is an increased risk for the loss of power system stability, i. e., stability will be lost if a severe disturbance occurs. If an emergency, corrective, control action is not performed, the system will go unstable. The response requirements (time and reliability) are normally higher for emergency control actions than for normal control actions. Emergency control functions are almost always automatic, while normal control actions can be either automatic or manual, e. g., in conjunction with alarms. The actions taken in the power system are, however, quite similar for both normal control and emergency control. Protection, system protection and emergency control comprise corrective (or curative) measures, i. e., actions are really needed to save the component or the system. Protection could very well be regarded as binary (on/off) emergency control, but by tradition, protection is quite specific.

Voltage control comprises actions like AVR, tap-changers and shunt devices, automatic or manual control. The control variable is usually the voltage level or reactive power flow. For system protection purposes or "emergency control," also load shedding and AGC can be used.

Primary frequency control is normally performed in the power stations by the governor controls, while secondary frequency control is performed by the AGC change of set-point or start/stop of units by the dispatcher. For emergency control, load shedding as well as actions that reduces the voltage, and hence the voltage-sensitive part of the load, can be used.

Automatic power flow control is performed by AGC, HVDC control, united power flow controller (UPFC), thyristor-controlled series capacitor (TCSC), phase-shift transformers, etc. Load level, network topology, and generation dispatch are the most common parameters that influence the power flow.

Angle control is more accurate if based on PMUs. Without PMUs, power flow is an indirect method of measuring and controlling the angle. The actions are similar as for power flow control.

SCADA/EMS functions are tools that assist the power system/grid operator in his effort to optimize the power system operation, with respect to economy, operational security, and robustness, as well as human and material safety. Operator actions are normally (at least supposed to be) preventive, i.e., actions taken to adjust the power system operational conditions to the present and near future expected situation. Preventive actions, based on simple criteria, can beneficially be implemented in SCADA/EMS and be performed automatically or be suggested and then released or blocked by the operator. SCADA/EMS systems are normally too slow to capture power system dynamics.

4.3.2 Power Generation Control

Fig. 4.8 shows a block diagram of a typical electric power plant. From a control standpoint, a large plant typically has a local, dedicated control system linked with a SCADA RTU via either a communications link or by discrete inputs and outputs. The local control system includes automatic voltage regulator (AVR), rotor speed controller and power system stabilizer (PSS). AVR controls the generator terminal voltage and keeps the generator terminal voltage output being equal to the reference voltage. Rotor speed controller makes the generator frequency being equal to 50 Hz or 60 Hz. PSS enhances the damping ratio of power system and damps the power swing.

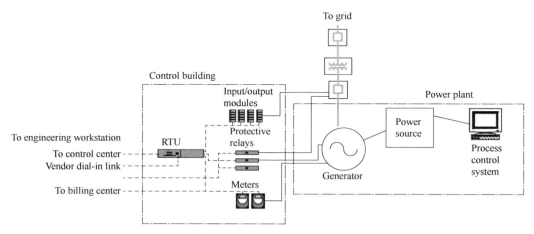

Fig. 4.8 Typical electric power plant block diagram

The SCADA RTU receives status and power flow data from the plant and, depending on the power plant, may send control signals to increase or decrease power output based on current load on the system and whether import/export of power from the system is required. **The generator typically has one or more protective relays to prevent damage to the generator or to the system. Newer** protective relays are microprocessor-based and one relay can incorporate all the functions necessary to protect the generator. With older electromechanical relays, several relays are required to perform the same function.

4.3.3 Power Substations Control

There are many similarities between power substations control at the transmission and substations at the distribution level, but a few differences as well. The following two sections describe their features, similarities, and differences.

1. Transmission Substations Control

Fig. 4.9 shows a diagram of a somewhat typical transmission substation control. Major power system components include circuit breakers, transformers, switches, and possibly capacitor banks. Major control and monitoring components include RTUs and their associated input/output modules, protective relays, and meters. Newer protective relays and meters are microprocessor-based and are often called IEDs.

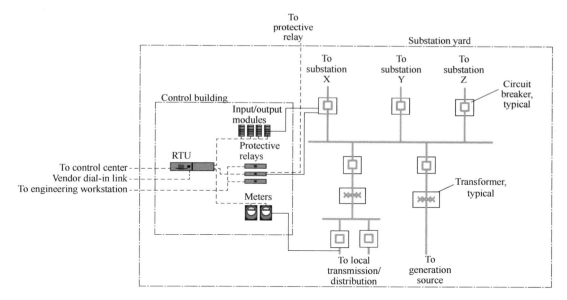

Fig. 4.9 Typical transmission substation control

RTUs interface with input modules to gather status data including circuit breaker open/closed status, transformer alarms, protective relay trip status, and other data. RTUs interface with outputs to control circuit breakers, switches, and transformer tap-changers. Depending on the size of the system and manufacturer, input/output modules can either be separate, standalone modules, or cards that slide into the RTU.

Protective relays function to prevent damage to major equipment in the event of an abnormal condition such as a short circuit. These relays monitor each transformer in the substation as well as lines going to other substations or generation plants. For lines going to other substations or plants, a relay is required at each end with communication between them so that if there is an abnormal condition both ends of the line can be opened to remove it from service and prevent damage to equipment. Communications channels can be via phone line, microwave, fiber optics, pilot wire, or power line carrier.

Newer relays, in addition to communicating with each other, have communications channels that are often tied into the RTU to gather metering data (line voltage, current, and power). They are also often tied into an engineering workstation so that engineers can access the relay in the event it trips a circuit breaker. The integration of protective relays into SCADA systems has raised the stakes somewhat with regard to cyber security. Utilities are much less likely to operate their system without protective relays and protection against damage to major, long-lead equipment. If an attacker breaks into a SCADA system and trips a breaker via an RTU, it can be closed again remotely. On the other hand, if an attacker could gain access to a protective relay to trip a breaker, the relay often triggers a lockout relay that prevents a breaker from being closed again until the lockout is reset at the substation.

Meters at the transmission level are used to determine energy flowing into and out of a substation. This is particularly important if the energy is flowing out of a utility's system into another utility's system or vice versa. Newer meters are connected to RTUs via a communications link; older meters are connected via analog channels.

2. Distribution Substations

Distribution substations are similar to transmission substations in that they still have circuit breakers and transformers as their primary electrical components. From a control standpoint, they still have RTUs, protective relays, and meters that perform basically the same functions as in transmission substations. However, there are a few differences. Fig. 4.10 shows a typical distribution substation control.

A distribution substation typically steps the incoming voltage from transmission levels down to medium voltage levels (2400-69000 V) for utilization. Since lines at this level typically feed loads directly, protective relays typically do not have a communications link to another relay. These relays are also more likely to be the older, electromechanical type with no communications capability.

In recent years, automated meter reading and wireless control of distribution switches have become cost effective. Many utilities employ one or both of these functions to reduce manpower requirements and streamline operations. Automated meter reading gathers billing information via either a wireless or power line carrier communications link. The information is then uploaded to the utility's billing center, which may or may not be collocated with the control center. Automated pole-top distribution switches are typically controlled by a substation RTU via a wireless communications link. According to a recent survey, the popularity of these switches is increasing. The survey shows that respondents had 3823 of these switches currently installed with plans to install 2272 more. The additional communications channels provide more pathways for potential attackers to exploit.

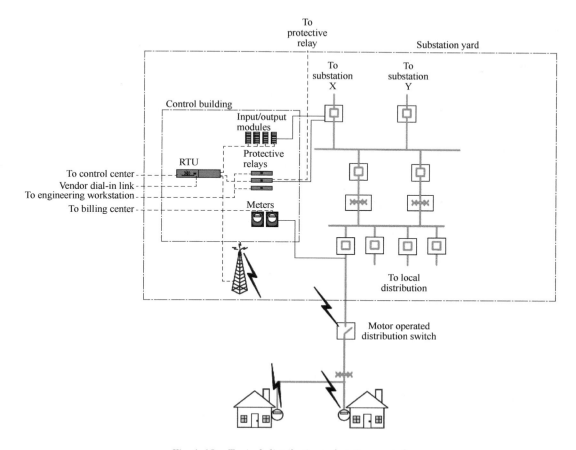

Fig. 4.10 Typical distribution substation control

4.3.4 Electric Utility Control Center

Key to the operation and maintenance of a modern utility's generation, transmission, and distribution resources as described in the previous sections is a centralized control and monitoring system. For most utilities, this system is housed in one or more control centers. The control center is the central nerve system of the power system. It senses the pulse of the power system, adjusts its condition, coordinates its movement, and provides defense against exogenous events.

Fig. 4.11 shows a diagram of a control center typical of a large utility. For utilities serving many thousands or millions of customers, several of these control centers may be used to monitor regional portions of the system. Smaller utilities and rural electric cooperatives typically have a subset of the equipment shown in Fig. 4.11. Components of the control center include the SCADA system, the energy management system, and other application servers and/or workstations.

A large control center typically is staffed by several operators. Each operator is often dedicated to a portion of the system such as transmission, distribution, or generation. The control center is often set up with separate areas for each of these functions as well.

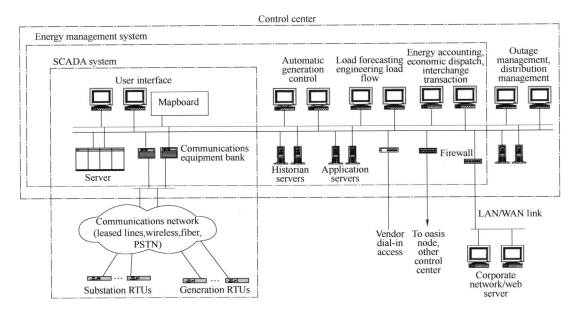

Fig. 4.11 Typical utility control center block diagram

1. SCADA System

SCADA is a term used in several industries fairly generically to refer to a centralized control and monitoring system. In the electric utility industry, SCADA usually refers to basic control and monitoring of field devices including breakers, switches, capacitors, reclosers, and transformers. As shown in Fig. 4.11, a SCADA system includes data collection computers at the control center and RTUs in the field that can collectively monitor and control anywhere from hundreds to tens of thousands of data points. It also includes a user interface that is typically monitored around the clock. The user interface, in addition to one or more computer displays, usually includes a mapboard or large group displays to provide an overview of system status.

Also included in the SCADA system are the communications channels required to transmit information back and forth from the central computer (s) to the RTUs. The physical media used to create these channels typically consist of leased lines, dedicated fiber, wireless (licensed microwave or unlicensed spread spectrum radio), or satellite links.

2. Energy Management System.

Most utilities have, in addition to a SCADA system, a computer system that coordinates and optimizes power generation and transmission. The system that performs this function is called an energy management system (EMS). As shown in Fig. 4.12, EMS can include applications such as AGC, load forecasting, engineering load flow, economic dispatch, energy accounting, interchange transaction, reserve calculations (spin and non-spin), and VAR/voltage control.

AGC controls generation units in real-time to maintain the system frequency at or very near 60 Hz. It also balances overall power generation with overall load. AGC is also used to import or export power from a utility's system. Increasing system frequency will cause power to be exported;

decreasing frequency causes power to be imported.

Load forecasting uses real-time data like outside temperature and historical data to predict the load hours or days in advance. Economic dispatch is concerned with determining which generators should be operated, based on system load and fuel costs, among other things. Interchange transaction manages the import and export of power from a utility's system. Reserve calculation compares actual generator output to rated output to determine reserve. Spinning reserve counts only those generators currently online. Non-spinning reserve includes those generators that are currently offline.

4.3.5 The Future of The Electric Utility Control Center

1. Introduction

At first, we give a brief historical account of the evolution of control centers. A great impetus to the development of control centers occurred after the northeast blackout of 1965 when the commission investigating the incident recommended that "utilities should intensify the pursuit of all opportunities to expand the effective use of computers in power system planning and operation. Control centers should be provided with a means for rapid checks on stable and safe capacity limits of system elements through the use of digital computers." The resulting computer-based control center, called EMS, achieved a quantum jump in terms of intelligence and application software capabilities. The requirements for data acquisition devices and systems, the associated communications, and the computational power within the control center were then stretched to the limits of what computer and communication technologies could offer at the time.

Special designed devices and proprietary systems had to be developed to fulfill power system application needs. Over the years, information technologies have progressed in leaps and bounds, while control centers, with their nonstandard legacy devices and systems that could not take full advantage of the new technologies, have remained far behind. Recent trends in industry deregulation have fundamentally changed the requirements of the control center and have exposed its weakness. Conventional control centers of the past were, by today's standards, too centralized, independent, inflexible, and closed.

The restructuring of the power industry has transformed its operation from centralized to coordinated decentralized decision-making. The blackouts of 2003 may spur another jump in the applications of modern information and communication technologies (ICT) in control centers to benefit reliable and efficient operations of power systems. The ICT world has moved toward distributed intelligent systems with web services and grid computing. The idea of grid computing was motivated by the electric grids of which their resources are shared and consumers are unaware of their origins. The marriage of grid computing and service-oriented architecture into grid services offers the ultimate decentralization, integration, flexibility, and openness. We envision a grid services-based future control center that is characterized by:

1) An ultrafast data acquisition system;
2) Greatly expanded applications;

3) Distributed data acquisition and data processing services;
4) Distributed control center applications expressed in terms of layers of services;
5) Partner grids of enterprise grids;
6) Dynamic sharing of computational resources of all intelligent devices;
7) Standard grid services architecture and tools to manage ICT resources.

Control centers today are in the transitional stage from the centralized architecture of yesterday to the distributed architecture of tomorrow. In the last decade or so, communication and computer communities have developed technologies that enable systems to be more decentralized, integrated, flexible, and open. Such technologies include communication network layered protocols, object technologies, middleware, etc. which are briefly reviewed in this sector. Control centers in power systems are gradually moving in the directions of applying these technologies.

The trends of present-day control centers are mostly migrating toward distributed control centers that are characterized by:
1) SCADA, EMS, and business management system (BMS);
2) IP-based distributed SCADA;
3) common information model (CIM)-compliant data models;
4) Middleware-based distributed EMS and BMS applications.

Control centers today, not surprisingly, span a wide range of architectures from the conventional system to the more distributed one described above.

2. Power System Control Center Functions

In the second half of the 1990s, a trend began to fundamentally change the electric power industry. This came to be known as industry restructuring or deregulation. Vertically integrated utilities were unbundled; generation and transmission were separated. Regulated monopolies were replaced by competitive generation markets. Transmission, however, remained largely regulated. The principle for the restructuring is the belief that a competitive market is more efficient in overall resource allocation. While suppliers maximize their profits and consumers choose the best pattern of consumption that they can afford, the price in a market will adjust itself to an equilibrium that is optimal for the social welfare.

Two types of markets exist in the restructured power industry. One is the bilateral contracts between suppliers and consumers. The other one is an auction market in which generators submit bids to a centralized agent which determines the winning bids and the price. The price could be determined by a uniform pricing scheme (in the United States) based on the highest bid price that is deployed to serve the load, or a nonuniform pricing scheme (in the U.K.) based on the bid price (pay-as-bid). A market operator is needed to run the auction market. The market operator may be an independent system operator (ISO), a regional transmission organization (RTO), or other entities with similar functions. With the introduction of electricity markets, some control centers, e.g., that of an ISO or an RTO, have had the responsibility of running the market operation, as well as maintaining system reliability.

The two aspects are usually separated but with close coordination. The electricity market is different from any other commodity market in that power has to be balanced at all times. This requirement leads to a more complex market structure. Most electricity markets have a day-ahead energy market, a real-time balancing market, and an ancillary service market. The day-ahead market is a forward market in which hourly clearing prices are calculated for each hour of the next day based on generation and demand bids, and bilateral transaction schedules. If the reliability of the transmission system imposes limits on most economical generation causing transmission congestion to occur, congestion management is required. One of the approaches to congestion management is based on the pricing differences, called locational marginal prices (LMP), between the nodes of the network. The LMPs are obtained from the nodal shadow prices of an optimal power flow (OPF) or security-constrained economic dispatch (SCED). The day-ahead market enables participants to purchase and sell energy at these binding day-ahead LMPs. A security-constrained unit commitment (SCUC) is conducted, based on the bids submitted by the generators, to schedule the startup and shutdown of generators in advance. The transmission customers may schedule bilateral transactions at binding day-ahead congestion charges based on the difference in LMPs at the generation and the demand sides. The balancing market is the real-time energy market in which the market clearing prices (a new set of LMPs) are calculated using SCED every 5 min or so based on revised generation bids and the actual operating condition from state estimation. Any amount of generation, load, or bilateral transaction that deviates from the day-ahead schedule will pay the balancing market LMP. There is a host of supporting functions to ensure reliable delivery of electricity to consumers. To ensure against possible generator failure and/or sudden load increase, additional generation capacity has to be provided. This real power reserve is always ready to take on the responsibility instantaneously. Adequate reactive power resources are needed to maintain voltage at an acceptable level for proper operation of the system. These are all grouped under the name of ancillary services. An ancillary service can either be self-provision by users of the transmission system or system-wide management by the ISO/RTO. Markets have also been established to manage ancillary services.

Industry restructuring has so far brought two major changes in control centers structures. The first one is the expansion of the control center functions from traditional energy management, primarily for reliability reasons, to business management in the market. The second one is the change from the monolithic control center of traditional utilities that differed only in size to a variety of control centers of ISOs or RTOs, transmission companies (TransCos), generation companies (GenCos), and load serving entities (LSEs) that differ in market functions. The control centers in the market environment are structured hierarchically in two levels, as shown in Fig. 4.12.

In Fig. 4.13, the ISO/RTO control center (CC) that operates the electricity market of the region coordinates the LSE and other control centers for system reliability in accordance with market requirements. All entities, ISO, RTO, LSE, GenCo, etc., are market participants. Their control centers are equipped with business functions to deal with the market. The part of control center functions that are responsible for business applications is BMS. The ISO or RTO is usually the market operator; therefore, its BMS is also called the market operations system (MOS). There are close

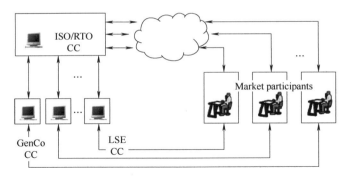

Fig. 4.12 Control centers (CC) in the market environment

interactions between the functions of EMS and BMS as shown in Fig. 4.13. For other types of control centers that do not operate a market, a BMS is added to the traditional EMS to interact with the market.

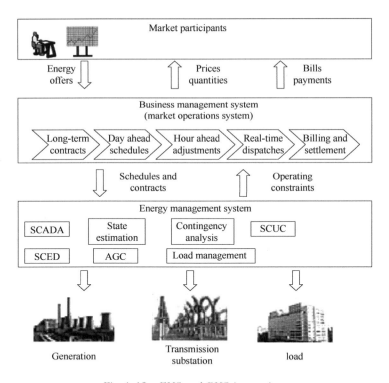

Fig. 4.13 EMS and BMS interactions

Control centers have evolved over the years into a complex communication, computation, and control system. The control center will be viewed here from functional and architectural perspectives. As pointed out previously, there are different types of control centers whose BMS are different. From the functional point of view, the BMS of the control center of ISO/RTO is more complex than the others. Our description below is based on a generic control center of ISO/RTO, whereas specific ones may be somewhat different in functions and structure.

From the viewpoint of the system's user, a control center fulfills certain functions in the operation of a power system. The implementations of these functions in the control center computers are, from the software point of view, called applications.

The first group of functions is for power system operation and largely inherits from the traditional EMS. They can be further grouped into data acquisition, generation control, and network (security) analysis and control. Typically, data acquisition function collects real-time measurements of voltage, current, real power, reactive power, breaker status, transformer taps, etc. from substation RTUs every 2 s to get a snapshot of the power system in steady-state. The collected data is stored in a real-time database for use by other applications. The sequence of events (SOE) recorder in an RTU is able to record more real-time data in finer granularity than they send out via the SCADA system. These data are used for possible post-disturbance analysis. Indeed, due to SCADA system limitations, there are more data bottled up in substations that would be useful in control center operations. Generation control essentially performs the role of balancing authority in NERC's functional model. Short-term load forecasts in 15 min intervals are carried out. AGC is used to balance power generation and load demand instantaneously in the system. Network security analysis and control, on the other hand, performs the role of reliability authority in NERC's functional model. State estimation is used to cleanse real-time data from SCADA and provide an accurate state of the system's current operation. A list of possible disturbances, or contingencies, such as generator and transmission line outages, is postulated and against each of them, power flow is calculated to check for possible overload or abnormal voltage in the system. This is called contingency analysis or security analysis.

The second group of functions is for business applications and is the BMS. For an ISO/RTO control center, it includes market clearing price determination, congestion management, financial management, and information management. Different market rules dictate how the market functions are designed. The determination of market clearing price starts from bid management. Bids are collected from market participants. A bid may consist of start-up cost, no-load cost, and incremental energy cost. Restrictions may be imposed on bids for market power mitigation. Market clearing prices are determined from the acceptable bids. SCUC may be used to implement day-ahead markets. In a market with uniform pricing or pay-as-bid, the determination is done simply by the stack-up of supply versus demand. If the LMP that incorporates congestion management is applied, an OPF or SCED will be used. Other congestion management schemes such as uplift charges shared by all for the additional charges resulting from congestion are employed in some markets. To manage the risk of congestion charge volatility, a hedging instrument called transmission right is introduced. Transmission rights can be physical or financial. Physical rights entitle the holder the rights to use a particular portion of the transmission capacity. Financial rights, on the other hand, provide the holder with financial benefits equal to the congestion rent. The allocation and management of transmission rights are part of market operations and require running OPF. Financial management functions in the electricity market include accounting and settlement of various charges.

Fig. 4.14 highlights some major functions of BMS and EMS in today's control centers. In the deregulated environment, the AGC set points and the transaction schedules are derived from BMS, or the MOS, instead of the traditional EMS ED and interchange scheduling (IS). The BMS uses the network model, telemetry data and operating constraints from EMS to clear the market. We will explain the other blocks (ERP and data warehouse) outside the dotted lines of EMS and BMS in Fig. 4. 14.

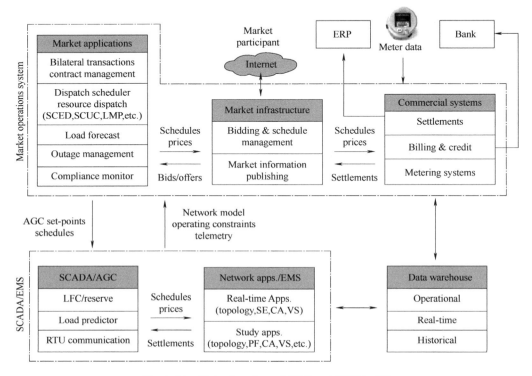

Fig. 4. 14 Major functions of BMS and EMS in CC

For a fair and transparent use of the transmission system, certain information needs to be available to the public and such information is posted, in the United States, through the Internet at the open access same-time information system (OASIS). Also, in compliance with NERC requirements, the tagging scheduling and checkout functions are used by control centers to process interchange transaction schedules.

3. Architecture of Power System Control Center

The SCADA system was designed at a time when the power industry was a vertically integrated monopoly. The centralized star configuration in which data from several remote devices were fed into a single computer was a ubiquitous configuration in the process control industry. This architecture fits the needs of the power system then. Over the years, networking and communications technologies in the computer industry have progressed significantly. But in the power industry they had not changed much and the SCADA system had served its needs well until the recent onset of deregulation. Substation automation in recent years, however, has introduced digital relays and other digital measurement devices; all called IEDs. A RTU could become another IED. The IEDs in a substation are linked by a LAN. The computer in the control center or the EMS, serving generation control and network analysis applications, has advanced from mainframes, to minis, to networks of workstations or PCs. A dual-configured LAN with workstations or PCs is commonly adopted in the control centers. The inter control center connections are enabled through point-to-point networks for data transfer. The BMS, on the other hand, communicate through the Internet. The control center thus has several networks: a star master-slave network from RTUs to the control center with dedicated physical links, a LAN for EMS application servers, a point-to-point network for inter control center connections, and the Internet for BMS market functions. The substation control center has a LAN for its SCADA and distribution feeder and other automation functions (Fig. 4. 15).

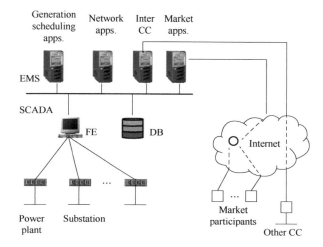

Fig. 4. 15 Substation CC architecture

Large amounts of data are involved in a control center. In a conventional control center, real-time data are collected from RTUs. Historical data and forecasted data are stored in storage devices. Different sets of application data are used by different application servers. Display file data are used by graphical user interface (GUI) workstations. Various copies of data have to be coordinated, synchronized, and merged in databases. Historically, proprietary data models and databases were used as a result of proprietary RTU protocols and proprietary application software to which the databases were attached. Power systems are complex; different models of the same generator or substation with varying degrees of granularity and diverse forms of representations are used in different applications, e. g., state estimation, power flows, or contingency-constrained economic dispatch.

4. Modern Distributed Control Centers

The term "distributed control center" was used in the past to refer to the control center whose applications are distributed among a number of computers in a LAN. By that standard almost all control centers today that are equipped with distributed processing capability in a networked environment would be called "distributed." That definition is too loose. On the other hand, if a control center that is fully decentralized, integrated, flexible, and open, is definited as a distributed control center. We adopt the definition of the distributed system to control centers and call it a distributed control center if it comprises of a set of independent computers that appears to the user as a single coherent system. A distributed control center typically has some or all of its data acquisition and data processing functions distributed among independent computers and its EMS and BMS applications also distributed. It utilizes the distributed system technologies to achieve some level of decentralization, integration, flexibility, and openness: the characteristics that is desirable in today's power system operating environment.

Current trends in the development of distributed control centers from the previous multicomputer networked system to a flexible and open system with independent computers are moving in the following directions:

1) Separation of SCADA, EMS, and BMS;
2) IP-based distributed SCADA;
3) Standard (CIM)—based distributed data processing;
4) Middleware—based distributed EMS and BMS applications.

The data acquisition part of SCADA handles real-time data, and is very transaction-intensive. The applications in EMS involve mostly complex engineering calculations and are very computation-intensive. These two dissimilar systems are tightly bundled together in a conventional control center because proprietary data models and databases are used due to historical reasons. With proprietary data models and database management systems to handle data, such data could not be easily exchanged, and it prevents effective use of third-party application software. A separate SCADA system and EMS system would serve a control center better by expediting the exploitation of new technologies to achieve the goals of decentralization, integration, flexibility, and openness. The separation of SCADA and EMS is a logical thing to do.

The SCADA function in a conventional control center starts with the RTUs collecting data from substations and, after simple local processing (e.g., data smoothing and protocol specification), the data is then sent through a dedicated communication channel with proprietary protocol to the appropriate data acquisition computer in the control center where a TCP/IP based computer network is used. An interface is therefore needed. The data acquisition computer converts the data and prepares it for deposition in the real-time database. The real-time database is accessed and used by various applications. For reasons of efficiency, the interface may be handled by a telecontrol gateway, and more gateways may be used in a control center. The location of the gateway may move to the substation if standard IP protocol is used. The gateway is then connected to the control center. If the RTU is TCP/IP based, it can be connected directly to the control center resulting in a distributed data acquisition system. In this way, the gateway serves as a data concentrator and communication processor (Fig. 4.16). RTUs or IEDs may be connected to a data concentrator or connected directly to the control center.

The use of fixed dedicated communication channels from RTUs to control center leaves no flexibility in RTU-control center relationship which was not a problem in the past. When, for example, another control center requires real-time data from a particular substation not in the

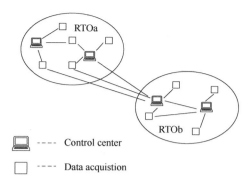

Fig. 4.16 Distributed data acquisition

original design, there are only two ways to do it. In the first case, the other control center has to acquire the data through the control center to which the RTU is attached. In the second case, a new dedicated channel has to be installed from the RTU to the other control center.

The dedicated channels with proprietary protocols for SCADA were developed for reasons of speed and security. Communication technologies have advanced tremendously in the last couple of decades in both hardware and software, resulting in orders of magnitude increase in transmission speed and sophistication in security protection. Modern communication channels with metallic or optical cables have enormous bandwidths compared to the traditional 9600 Kb/s or less available for RTU communications. Responsibility to guarantee timely delivery of specific data in a shared communication network such as intranet or Internet falls to the QoS function of communication network management and is accomplished through protocols for resource allocation. With further advancement in QoS, more and more stringent real-time data requirements may be handled through standard protocols. Network security involves several issues: confidentiality, integrity, authentication, and nonrepudiation. Cryptography is used to ensure confidentiality and integrity, so that the information is not available to and can not be created or modified by unauthorized parties. Digital hash is used for authentication and digital signature for nonrepudiation. Network security has advanced rapidly in recent years.

There is no reason that the new SCADA communications should not be using standard protocols such as IP-based protocols. Indeed, inability of SCADA to take advantage of recent progress in cyber security is considered a serious security risk by today's standards. SCADA should be IP-based and liberated from dedicated lines by tapping into an available enterprise WAN or as a minimum use internet technology to enable the use of heterogeneous components. In the future, as internet QoS performance and encryption protocols improve, there will be little difference between a private-line network and virtual private network (VPN) on the internet when standard protocols are used. A VPN is a network that is constructed by using public wires to connect nodes. The system uses encryption and other security mechanisms to ensure that only authorized users can access the network and that the data cannot be intercepted. VPN can be used to augment a private-line enterprise network when a dedicated facility can not be justified. In other words, the physical media and the facilities used for the network will become less of a concern in a control center when standard protocols are used.

If standard communication protocols are used, a control center (i.e., its application software) may take data input from data concentrators situated either inside or outside its territory. The only difference between data from an internal data concentrator or an external one is the physical layer of the communication channel. The former may be through the intranet and the latter needs special arrangement for a communication channel (Fig. 4.17).

With IP-based distributed SCADA that uses standard data models, data can be collected and processed locally before serving the applications. Again using standard data models, databases can also be distributed. Search engines may be utilized for fast and easy access to relevant information. Intelligent agents with learning capability may be deployed for data management and data delivery. Once the data is deposited in the real-time database, it can be used by various applications to serve the required functions of EMS or BMS. The output of an application may be used by another application. As long as an application has access through the network to the database with sufficient speed and ensured security, the physical location of the application server and the data will be of little concern. The network should be capable of hosting any kind of applications and supporting intelligent information gathering through it. Component and middleware technologies enable such distributed architecture.

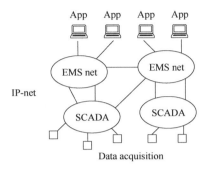

Fig. 4.17 IP-based distributed SCADA

Present-day control centers are mostly provided with CIM data models and middleware that allow distributed applications within the control center. Only a few of them use CIM-based data models and middleware-based applications as their platforms. Specific applications of distributed technologies include Java, component technology, middleware-based distributed systems and agent technology.

A comprehensive autonomous distributed approach to power system operation is proposed. Deployment of agents responsible for specific temporally coordinated actions at specific hierarchical levels and locations of the power system is expected to provide the degree of robust operation necessary for realizing a self-healing grid.

5. Future Control Centers

Future control centers, as we envision, will have much expanded applications both in power system operations and business operations based on data that are collected in a much wider and faster scale. The infrastructure of the control center will consist of large number of computers and embedded processors (e.g., IEDs) scattered throughout the system, and a flexible communication network in which computers and embedded processors interact with each other using standard interfaces. The data and data processing, as well as applications, will be distributed and allow local and global cooperative processing. It will be a distributed system where locations of hardware, software, and data are transparent to the user.

The information technology has evolved from objects, components, to middleware to facilitate the development of distributed systems that are decentralized, integrated, flexible, and open. Although significant progress has been made, it is still not fully distributed. Today's middleware is somewhat tightly coupled. Recent progress in the ease and popularity with XML-based protocols in internet applications has prompted the development of the web services architecture, which is a vital step toward the creation of a fully distributed system. The concept of services represents a new paradigm. A software service was originally defined as an application that can be accessed through a programmable interface. Services are dynamically composed and distributed, and can be located, utilized, and shared. The developers in grid computing have extended the concept of service from software to resources such as CPU, memory, etc. Resources in grid computing can be dynamically composed and distributed, and can be located, utilized, and shared. Computing, as a resource service, is thus distributed and shared. The idea of service-oriented architecture, grid computing and open standards should be embraced for adoptions not only in future control centers, but also in other power system functions involving information and communication technologies.

In a grid services environment, data and application services, as well as resource services, are distributed throughout the system. The physical location of these services will be of little concern. It is the design of the specific function in an enterprise that dictates how various services are utilized to achieve a specific goal. The control center function, i.e., to ensure the reliability of power system operation and to manage the efficient operation of the market, represents one such function in the enterprise. In this new environment, the physical boundary of any enterprise function, such as the control center, may no longer be important and indeed become fuzzy. It is the collective functionality of the applications that represents the control center makes it a distinct entity.

Grid services-based future control centers will have distributed data services and application services developed to fulfill the role of control centers in enhancing operational reliability and efficiency of power systems. Data service will provide just-in-time delivery of information to applications that perform functions for power system control or business management. The computer and communication infrastructure of future control centers should adopt standard Grid services for management of the resources scattered around the computers and embedded processors in the power system to support the data and application services.

The concepts we just brought up about future control centers: extended data acquisition, expanded applications, web services, grid computing, etc., will be elaborated in more details in the remaining subsections.

(1) **Data Acquisition**

For power system reliability, the security monitoring and control function of the control center is actually the second line of defense. The first line of defense is provided by the protective relay system. For example, when a fault in the form of a short circuit on a transmission line or a bus occurs, measurement devices such as a current transformer (CT) or potential transformer (PT) pick up the information and send it to a relay to initiate the tripping (i.e., opening) of the appropriate circuit breaker or breakers to isolate the fault. The protective relay system acts in a matter of one or two cycles (one cycle is 1/60 of a second in a 60 Hz system). The operation of the protective relay system is based on local measurements. The operation of security monitoring and control in a control center, on the other hand, is based on system-wide (or wide-area) measurements every 2 s or so from the SCADA system. The state estimator in EMS then provides a snapshot of the whole system. Different time scales driving the separate and independent actions by the protective system and the control center lead to an information and control gap between the two. This gap has contributed to the missed opportunity in preventing cascading outages, such as the North American blackout and the Italian blackout of 2003. In both cases, protective relays operated according to their designs by responding to local measurement, whereas the control center did not have the system-wide overall picture of events unfolding. During that period of more than half an hour, control actions could have been taken to save the system from a large-scale blackout.

The security monitoring and control functions of today's control center, such as state estimation, contingency analysis, etc., are based on steady-state models of the power system. There is no representation of system dynamics that govern the stability of a system after a fault in control center's advanced application software. The design philosophy for security control is that of preventive control, i. e., changing system operating conditions before a fault happens to ensure the system can withstand the fault. There is no analytical tool for emergency control by a system operator in a control center. All of these are the result of limitations imposed by the data acquisition system and computational power in conventional control centers. The issue of computational power has already been addressed by grid computing. We will discuss the issue of measurement system in what follows.

Although RTUs, IEDs, and substation control systems (SCSs) in substations sample power measurements in a granularity finer than a second, SCADA collects and reports data (by exception) only in the interval of several seconds. The system-wide measurements, strictly speaking, are not really synchronized, but their differences are in the order of the time-scale of the window of data collection, which is approximately 2 s. But because the model is a steady-state model of the power system, such a discrepancy is tolerated. As mentioned earlier, the bandwidth limitation problem in SCADA is a legacy problem from the past. Future communication networks for SCADA using WAN will have much wider bandwidth and will be able to transmit measurement data in finer resolutions. However, the data needs to be synchronized.

This can be done by using synchronization signals from the global positioning system (GPS) via satellites. Modern GPS-based phasor measurement units (PMU) are deployed in many power systems to measure current and voltage phasors and phase angle differences in real-time. GPS in PMU provides a time-tagged one pulse-per-second (PPS) signal which is typically divided by a phase-locked oscillator into the required number of pulses per second for sampling of the analog measurements. In most systems being used at present, this is 12 times per cycle or 1.4 ms in a 60 Hz system. In principle, system-wide synchronous data in the order of milliseconds or even finer can be collected by PMU-like devices in the future to be used for monitoring system dynamic behavior. PMUs and the future generation of PMU-class data acquisition devices can augment existing RTUs, IEDs, and SCSs to provide a complete picture of power system dynamical conditions and close the gap between today's protective relay operations and control center functions.

The electricity market is a new experiment. As the market matures and our understanding of market operation strengthens, new measurement requirements and devices, and new market information or data acquisition systems that will ensure an efficient and fair market will definitely emerge. Real-time measurements are needed and techniques should be developed for market surveillance to mitigate market power and for enforcement of contract compliance. Real-time data, properly collected and judiciously shared, can also assist regional cooperation, such as regional relay coordination or regional transmission planning among interconnected systems, that benefits all parties involved.

(2) Functions

Market functions of a control center will expand once new measurement systems are available and our understanding of market behavior increases. We mentioned market surveillance and contract compliance and there will be more in the future. On the power system operation side, the new data acquisition systems, such as PMUs that provide measurements in the order of milliseconds, offer new opportunities for dynamic security assessment and emergency control that would greatly enhance system reliability. A great deal of research has already begun along these directions.

Current control centers provide analysis and control of power systems based on the steady-state models of the power system. For system dynamic effects, such as transient stability, the approach has always been to conduct simulation studies based on postulated future conditions and the results are used to design protective system response and to set operational limits on transmission lines and other apparatuses. This approach is becoming more and more difficult to continue due to increasing uncertainty in system operation conditions in the market environment. Online monitoring and analysis of power system dynamics using real-time data several times a cycle will make it possible for appropriate control actions to mitigate transient stability problems in a more effective and efficient fashion. Other system dynamic performance, including voltage stability and frequency stability, can also be improved with the assistance of PMUs. A comprehensive "self-healing power grid" framework for coordinating information and control actions over multiple time-scales ranging from milliseconds to an hour employing distributed autonomous intelligent agents.

Another function in control centers that has developed rapidly in the last couple of years and will become even more important in the future is the visualization tools, to assist power system operators to quickly comprehend the "big picture" of the system operating condition. As technology progresses, more and more data will become available in real-time. The human-machine aspect of making useful information out of such data in graphics to assist operators comprehend the fast changing conditions easily and timely and be able to respond effectively is crucial in a complex system such as the power system as long as human operators are still involved. Developing new functions to utilize enhanced data acquisition systems to greatly improve power system reliability and efficiency will be a great challenge to the research community. Successful research results will be valuable in bringing power system operations to a new level of reliability and efficiency.

(3) Grid Services-Based Future Control Centers

A future grid service-based control center will be an ultimate distributed control center that is decentralized, integrated, flexible, and open. In a grid-service environment, everything is a service. Future control centers will have data services provided throughout the power system. Data acquisition services collect and timestamp the data, validate and normalize them, and then make it available. Data processing services process data from various sources for deposition into databases or high level applications. Applications will call data services and data will be delivered just in-time for critical applications. Various functions serving the needs of control centers are carried out as application services. Traditional applications, such as contingency analysis, congestion management, may be further decomposed into their constituent components, for example, power flows, OPF, etc. Application services may have different granularity and may rely on other services to accomplish its job (Fig. 4.18).

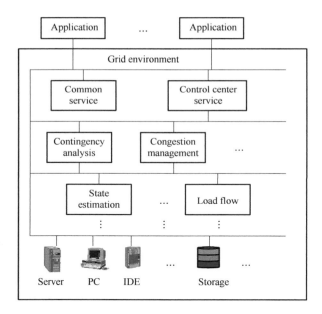

Fig. 4.18 Control center application services

Data and application services are distributed over the grids. The grids can use the intranet/internet infrastructure in which sub-networks are formed for different companies (enterprise grids) with relatively loose connection among cooperative companies (partner grid). The computer and communication resources in the grids are provided and managed by the standard resource services that deliver distributed computing and communication needs of the data and application services.

The designer of control centers develops data and application services, and no longer needs to be concerned with the details of implementation, such as the location of resources and information security, provided the services are properly registered in the grid environment. The new business model is that the software vendors will be service providers and power companies as service integrators. Power companies focus on information consumption and vendors focus on software manufacturing, maintenance, and upgrading. Computer and communication infrastructure will be left to the ICT professionals. This clear separation of responsibility would simplify and accelerate the delivery of new technology.

We envision future control centers based on the concept of grid services include among others the following features:

1) An ultrafast data acquisition system;
2) Greatly expanded applications;
3) A partner grid of enterprise grids;
4) Dynamic sharing of computational resources of all intelligent devices;
5) Use of service-oriented architecture;
6) Distributed data acquisition and data processing services;
7) Distributed control center applications expressed in terms of layers of services;
8) Use of standard grid services architecture and tools to manage ICT resources.

VOCABULARY

1. conventional   adj. 符合习俗的，传统的；常见的；惯例的
2. attenuator   n. 衰减器
3. attenuate   adj. 减弱的；稀薄的   vi. 变纤细；变弱
4. spike   n. 尖峰信号
5. optical   adj. 光学的；眼睛的，视觉的
6. cable   n. 电缆；海底电报
7. dynamic   n. 动态；动力   adj. 动态的；动力的；动力学的；有活力的
8. harmonic   n. [物]谐波；和声   adj. 和声的；谐和的；音乐般的
9. infrastructure   n. 基础设施；公共建设；下部构造
10. recloser   n. 重合闸装置
11. dispatch   n. 调度，派遣；急件   vt. 派遣；分派
12. surrogate   adj. 代理的；替代的   n. 代理；代用品；法官   vt. 代理
13. optimality   n. 最佳性
14. penalty   n. 罚款，罚金；处罚
15. compensate   vt. 补偿，赔偿；付报酬   vi. 补偿，赔偿；抵消
16. formula   n. 公式，准则；配方
17. contingency   n. 意外事故；偶然性；可能性
18. laymen   n. 非专业人员；外行
19. algorithm   n. 算法，运算法则
20. optimal   adj. 最佳的；最理想的
21. vendor   n. 卖主；供应商
22. feeder   n. 支线；馈电线；馈线
23. acronym   n. 首字母缩略词
24. refine   vt. 精炼，提纯；改善
25. valve   n. 阀门
26. dose   n. 剂量；一剂，一服   vt. 给药；配剂   vi. 服药
27. compatible   adj. 兼容的；能共处的；可并立的
28. pneumatic   n. 气胎；空气泵
29. telemetry   n. 遥测技术；遥感勘测；自动测量记录传导
30. blur   vt. 涂污；使…模糊不清   vi. 沾上

污迹；变模糊

31. lease *vt.* 出租；租得 *vi.* 出租
32. deployment *n.* 调度，部署
33. mimic *vt.* 模仿，模拟
34. hydraulic *adj.* 水力的；液压的；水力学的
35. pervade *vt.* 遍及；弥漫
36. monolithic *n.* 单块集成电路，单片电路
37. proprietary *adj.* 专有的，所有的；专利的；私人拥有的
38. protocol *n.* 协议；草案；礼仪
39. backplane *n.* 基架；背板，底板
40. multiple *adj.* 多样的；许多的；多重的 *n.* 并联；倍数
41. depict *vt.* 描述；描画
42. peripheral *adj.* 外围的；次要的
43. utilize *vt.* 利用
44. synchronization *n.* 同步；同时性
45. shunt *n.* 分流器；并联；转轨
46. grid *n.* 网格；格子，栅格；输电网
47. damp *vi.* 减幅，阻尼
48. stake *n.* 赌注；奖金；风险
49. exogenous *adj.* 外生的；外因的；外成的
50. spin *vi.* 旋转；晕眩
51. ancillary *adj.* 辅助的；副的；从属的 *n.* 助手；附件
52. granularity *n.* 间隔尺寸；粒度
53. coherent *adj.* 连贯的，一致的；凝聚性的
54. cryptography *n.* 密码学；密码使用法
55. heterogeneous *adj.* 多相的；异种的
56. encryption *n.* 加密；加密术
57. intercept *vt.* 拦截；截断 *n.* 拦截；截距
58. vital *adj.* 至关重要的；有活力的
59. surveillance *n.* 监督；监视

NOTES

1. Traditional measurement and protection systems consist of current and voltage transformers (CTs and PTs) that produce analog signals that are transported through electrical cables to the substation control room.

传统的测量和保护系统包含电流和电压变换器（CT 和 PT），二者产生的模拟信号通过电缆传送到变电站控制室。

2. The signal-conditioning circuit has four stages and provides the following functions：attenuating signals, suppressing voltage spikes, preventing ground loops, and filtering high-frequency noise over 2000 Hz.

信号调理电路包括四个部分并具备以下功能：衰减信号，抑制电压峰值，防止接地回路，滤除超过 2000 Hz 的高频噪声。

3. When digital computers were introduced in the 1960s, remote terminal units (RTUs) were developed to collect real-time measurements of voltage, real and reactive powers, and status of circuit breakers at transmission substations through dedicated transmission channels to a central computer equipped with the capability to perform necessary calculation for automatic generation control (AGC), which is a combination of LFC and ED.

在 20 世纪 60 年代引入数字化计算机时，开发了远程终端单元（RTUs）用于收集电压、有功与无功实时测量值及监测变电站断路器的状态，并将所收集信息通过专用传输通道上传

到配备有自动发电控制（AGC）计算执行能力的中央计算机。AGC 是负荷频率控制（LFC）和经济调度（ED）的结合。

4. These systems can be relatively simple, such as one that monitors environmental conditions of a small office building, or very complex, such as a system that monitors all the activity in a nuclear power plant or the activity of a municipal water system.

这些系统可以相对简单，如一个小型办公楼的环境状况监测系统；也可以非常复杂，如监测一个核电站的所有生产活动或一个市政水系统的生产活动。

5. At the same time, traditional PLCs included communications modules that would allow PLCs to report the state of the control program to a computer plugged into the PLC or to a remote computer via a telephone line.

同时，包含通信模块的传统 PLC 允许 PLC 把控制程序的状态报告给植入 PLC 的计算机或通过电话线传给远程计算机。

6. The central host computer or master station is most often a single computer or a network of computer servers that provide a man-machine operator interface to the SCADA system.

中央主机或主站通常是一台计算机或计算机服务器网络，它可以给 SCADA 系统提供一个人机操作界面。

7. Rather than the simple primary/standby failover scheme that was utilized in many first generation systems, the distributed architecture often kept all stations on the LAN in an online state all of the time.

与许多第一代系统经常使用的简单主/备用故障方案不同的是，分布式结构经常将局域网的操作工作站一直保持在线状态。

8. One set forms the master protocol, containing the valid statements for master station initiation or response, and the other set is the RTU protocol, containing the valid statements an RTU can initiate and respond to.

一个集合形成主站协议，包括主站的初始有效状态或响应有效状态。另一集合是 RTU 协议，包含 RTU 的初始有效状态或响应有效状态。

9. Since it is a protection system, it will operate in such operational situations when the power system would break down if no protective actions were taken.

由于这是一套保护系统，当电力系统即将崩溃时，如果保护没有动作，该保护系统将会起作用。

10. For emergency control, load shedding as well as actions that reduces the voltage, and hence the voltage-sensitive part of the load, can be used.

紧急控制下，在电压下降时采取甩负荷的措施，这样对电压敏感的那部分负荷就可以继续运行。

11. The SCADA RTU receives status and power flow data from the plant and, depending on the power plant, may send control signals to increase or decrease power output based on current load on the system and whether import/export of power from the system is required.

SCADA RTU 接收电厂的状态和潮流数据，并依据电厂情况，根据当前系统电流负荷和系统需要吸收还是输出功率，发送控制信号以增加或减少电厂的功率输出。

12. Key to the operation and maintenance of a modern utility's generation, transmission, and distribution resources as described in the previous sections is a centralized control and monitoring system.

运行和维护如前所述的现代电力公司发电、输电和配电资源的关键是集中控制和监测系统。

13. The price could be determined by a uniform pricing scheme (in the United States) based on the highest bid price that is deployed to serve the load, or a nonuniform pricing scheme (in the U.K.) based on the bid price (pay-as-bid).

价格可以由基于负荷售电最高投标价格的统一定价方案（美国）确定，或者由基于投标价格（按报价结算）的非统一定价方案（英国）确定。

14. As long as an application has access through the network to the database with sufficient speed and ensured security, the physical location of the application server and the data will be of little concern.

只要应用程序在通过网络访问数据库时有足够的速度并确保安全性，应用服务器和数据的物理位置将无关紧要。

15. Real-time data, properly collected and judiciously shared, can also assist regional cooperation, such as regional relay coordination or regional transmission planning among interconnected systems, that benefits all parties involved.

适当的收集和明智的共享实时数据，可以协助区域合作使参与各方受益，如跨网区域中继协调或区域输电规划等。

Chapter 5
High Voltage Direct Current (HVDC) Transmission

5.1 Introduction

The transmission system is becoming even more complex. High voltage (in either AC or DC electrical power transmission applications) is used for electric power transmission to reduce the energy lost in the resistance of the wires. For a given quantity of power transmitted and size of conductor, doubling the voltage will deliver the same power at only half the current. Since the power lost as heat in the wires is proportional to the square of the current, but does not depend in any major way on the voltage delivered by the power line, doubling the voltage in a power system reduces the line-loss, loss per unit of electrical power delivered by a factor of 4. Power loss in transmission lines can also be reduced by reducing resistance, for example, by increasing the diameter of the conductor, but larger conductors are heavier and more expensive.

High voltages cannot easily be used for lighting and motors, and so transmission-level voltages must be reduced to values compatible with end-use equipment. Transformers are used to change the voltage level in AC transmission circuits. The competition between the DC of Thomas Edison and the AC of Nikola Tesla and George Westinghouse was known as the War of Currents.

The advantage of HVDC is the ability to transmit large amounts of power over long distances with lower capital costs and with lower losses than AC. Depending on voltage level and construction details, losses are quoted as about 3% per 1000 km. High-voltage direct current transmission allows efficient use of energy sources remote from load centers.

In a number of applications HVDC is more effective than AC transmission. Examples include:

1) Undersea cables, where high capacitance causes additional AC losses. (e.g., 250 km Baltic Cable between Sweden and Germany, the 600 km NorNed cable between Norway and the Netherlands, and 290 km Basslink between the Australian mainland and Tasmania).

2) Endpoint-to-endpoint long-haul bulk power transmission without intermediate 'taps', for example, in remote areas.

3) Increasing the capacity of an existing power grid in situations where additional wires are difficult or expensive to install.

4) Power transmission and stabilization between unsynchronised AC distribution systems.

5) Connecting a remote generating plant to the distribution grid, for example Nelson River

Bipole.

6) Stabilizing a predominantly AC power-grid, without increasing prospective short circuit current.

7) Reducing line cost. HVDC needs fewer conductors as there is no need to support multiple phases. Also, thinner conductors can be used since HVDC does not suffer from the skin effect.

8) Facilitate power transmission between different countries that use AC at differing voltages and/or frequencies.

9) Synchronize AC produced by renewable energy sources.

Long undersea/underground high voltage cables have a high electrical capacitance, since the conductors are surrounded by a relatively thin layer of insulation and a metal sheath while the extensive length of the cable multiplies the area between the conductors. The geometry is that of a long co-axial capacitor. Where alternating current is used for cable transmission, this capacitance appears in parallel with load. Additional current must flow in the cable to charge the cable capacitance, which generates additional losses in the conductors of the cable. Additionally, there is a dielectric loss component in the material of the cable insulation, which consumes power.

However, when direct current is used, the cable capacitance is charged only when the cable is first energized or when the voltage is changed; there is no steady-state additional current required. For a long AC undersea cable, the entire current-carrying capacity of the conductor could be used to supply the charging current alone. The cable capacitance issue limits the length and power carrying capacity of AC cables. DC cables have no such limitation, and are essentially bound by only Ohm's Law. Although some DC leakage current continues to flow through the dielectric insulators, this is very small compared to the cable rating and much less than AC transmission cables.

HVDC can carry more power per conductor because, for a given power rating, the constant voltage in a DC line is the same as the peak voltage in an AC line. The power delivered in an AC system is defined by the root mean square (RMS) of an AC voltage, but RMS is only about 71% of the peak voltage. The peak voltage of AC determines the actual insulation thickness and conductor spacing. Because DC operates at a constant maximum voltage, this allows existing transmission line corridors with equally sized conductors and insulation to carry more power into an area of high power consumption than AC, which can lower costs.

Because HVDC allows power transmission between unsynchronized AC distribution systems, it can help increase system stability, by preventing cascading failures from propagating from one part of a wider power transmission grid to another. Changes in load that would cause portions of an AC network to become unsynchronized and separate would not similarly affect a DC link, and the power flow through the DC link would tend to stabilize the AC network. The magnitude and direction of power flow through a DC link can be directly commanded, and changed as needed to support the AC networks at either end of the DC link. This has caused many power system operators to contemplate wider use of HVDC technology for its stability benefits alone.

HVDC is less reliable and has lower availability than AC systems, mainly due to the extra conversion equipment. Single pole systems have availability of about 98.5%, with about a third of the downtime unscheduled due to faults. Fault redundant bipole systems provide high availability for 50% of the link capacity, but availability of the full capacity is about 97% to 98%.

The required static inverters are expensive and have limited overload capacity. At smaller transmission distances the losses in the static inverters may be bigger than in an AC transmission line. The cost of the inverters may not be offset by reductions in line construction cost and lower line loss. With two exceptions, all former mercury rectifiers worldwide have been dismantled or replaced by thyristor units. Pole 1 of the HVDC scheme between the North and South Islands of New Zealand still uses mercury arc rectifiers, as does Pole 1 of the Vancouver Island link in Canada. Both are currently being replaced—in New Zealand by a new thyristor pole and in Canada by a three-phase AC link. Efficient designs use silicon-controlled rectifiers (SCRs) (the more common name for thyristors) fired in sequence at 60 Hz to produce a modified sinewave of AC current, similar to the inverter circuitry in modern battery-operated UPSs for computer and telecom use.

In contrast to AC systems, realizing multiterminal systems is complex, as is expanding existing schemes to multiterminal systems. Controlling power flow in a multiterminal DC system requires good communication between all the terminals; power flow must be actively regulated by the inverter control system instead of the inherent impedance and phase angle properties of the transmission line. Multi-terminal lines are rare. One is in operation at the Hydro Québec—New England transmission from Radisson to Sandy Pond. Another example is the Sardinia-mainland Italy link which was modified in 1989 also provide power to the island of Corsica.

High voltage DC circuit breakers are difficult to build because some mechanism must be included in the circuit breaker to force current to zero, otherwise arcing and contact wear would be too great to allow reliable switching.

Operating a HVDC scheme requires many spare parts to be kept, often exclusively for one system as HVDC systems are less standardized than AC systems and technology changes faster.

Normally manufacturers such as Alstom, Siemens and ABB do not state specific cost information of a particular project since this is a commercial matter between the manufacturer and the client.

Costs vary widely depending on the specifics of the project such as power rating, circuit length, overhead vs. underwater route, land costs, and AC network improvements required at either terminal. A detailed evaluation of DC vs. AC cost may be required where there is no clear technical advantage to DC alone and only economics drives the selection.

However some practitioners have given out some information that can be reasonably well relied upon:

For an 8 GW 40 km link laid under the English Channel, the following are approximate primary equipment costs for a 2000 MW 500 kV bipolar conventional HVDC link (exclude way-leaving, on-shore reinforcement works, consenting, engineering, insurance, etc.):

1) Converter stations: £110M;
2) Subsea cable + installation: £1M/km.

So for an 8 GW capacity between England and France in four links, little is left over from £750M for the installed works. Add another £200-300M for the other works depending on additional onshore works required.

An April, 2010 announcement for a 2000 MW line, 64 km, between Spain and France, is 700 million euros, this includes the cost of a tunnel through the Pyrenees.

5.2 Converters in HVDC

5.2.1 Solid-State Power Device in HVDC

Power electronics can be defined as the conversion and conditioning of electric power using electronic switches. The five major types of power semiconductors used in solid-state AC motor control are:

1) Diodes;
2) Thyristors (e.g., SCR);
3) Transistors;
4) Gate-turn-off thyristors.

The commonality of these devices is the use of crystals of silicon in the form of wafers that are layered so as to form various combinations of P-N junction. The P junction is usually called the anode and N junction is usually called the cathode for diodes, SCRs, and GTOs; the corresponding terms for transistor are collector and emitter. The differences among these devices relate to how they go into and out of conduction and in their available ampere and voltage capabilities.

Most of the HVDC systems in operation today are based on line-commutated converters. Early static systems used mercury arc rectifiers, which were unreliable. Two HVDC systems using mercury arc rectifiers are still in service. The thyristor valve was first used in HVDC systems in the 1960s. The thyristor is a solid-state semiconductor device similar to the diode, but with an extra control terminal that is used to switch the device on at a particular instant during the AC cycle. The insulated-gate bipolar transistor (IGBT) is now also used, forming a voltage sourced converter, and offers simpler control, reduced harmonics and reduced valve cost. Because the voltages in HVDC systems, up to 800 kV in some cases, exceed the breakdown voltages of the semiconductor devices, HVDC converters are built using large numbers of semiconductors in series.

The low-voltage control circuits used to switch the thyristors on and off need to be isolated from the high voltages present on the transmission lines. This is usually done optically. In a hybrid control system, the low-voltage control electronics sends light pulses along optical fibres to the high-side control electronics. Another system, called direct light triggering, dispenses with the high-side electronics, instead using light pulses from the control electronics to switch light-triggered thyristors (LTTs). A complete switching element is commonly referred to as a valve, irrespective of its construction.

5.2.2 Rectifying Systems in HVDC Systems

Rectification and inversion use essentially the same machinery. Many substations (converter stations) are set up in such a way that they can act as both rectifiers and inverters. At the AC end, a set of transformers, often three-physically separated single-phase transformers, isolate the station from the AC supply, to provide a local earth, and to ensure the correct eventual DC voltage. The output of these transformers is then connected to a bridge rectifier formed by a number of valves. The basic configuration uses six valves, connecting each of the three-phases to each of the two DC rails. However, with a phase change only every sixty degrees, considerable harmonics remain on the DC rails.

An enhancement of this configuration uses 12 valves (often known as a twelve-pulse system). The AC is split into two separate three-phase supplies before transformation. One of the sets of supplies is then configured to have a star (wye) secondary, the other a delta secondary, establishing a 30° phase difference between the two sets of three-phases. With 12 valves connecting each of the two sets of three phases to the two DC rails, there is a phase change every 30°, and harmonics are considerably reduced.

In addition to the conversion transformers and valve-sets, various passive resistive and reactive components help filter harmonics out of the DC rails. Rectifier circuits are often classified by the number of current pulses that flow to the DC side of the rectifier per cycle of AC input voltage. A single-phase half-wave rectifier is a one-pulse circuit and a single-phase full-wave rectifier is a two-pulse circuit. A three-phase half-wave rectifier is a three-pulse circuit and a three-phase full-wave rectifier is a six-pulse circuit.

With three-phase rectifiers, two or more rectifiers are sometimes connected in series or parallel to obtain higher voltage or current ratings. The rectifier inputs are supplied from special transformers that provide phase shifted outputs. This has the effect of phase multiplication. Six phases are obtained from two transformers, twelve phases from three transformers and so on. The associated rectifier circuits are 12-pulse rectifiers, 18-pulse rectifiers and so on.

When controlled rectifier circuits are operated in the inversion mode, they would be classified by pulse number also. Rectifier circuits that have a higher pulse number have reduced harmonic content in the AC input current and reduced ripple in the DC output voltage. In the inversion mode, circuits that have a higher pulse number have lower harmonic content in the AC output voltage waveform.

1. Rectifier Devices

Before the development of silicon semiconductor rectifiers, vacuum tube diodes and copperoxide or selenium rectifier stacks were used. High power rectifiers, such as are used in high-voltage direct current power transmission, now uniformly employ silicon semiconductor devices of various types. These are thyristors or other controlled switching solid-state switches which effectively function as diodes to pass current in only one direction.

2. Half-Wave Rectification

In half-wave rectification, either the positive or negative half of the AC wave is passed, while the other half is blocked. Because only one half of the input waveform reaches the output, it is very inefficient if using power transfer. Half-wave rectification can be achieved with a single diode in a one-phase supply (Fig. 5.1), or with three diodes in a three-phase supply.

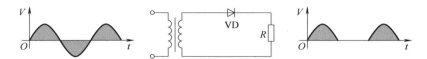

Fig. 5.1　A Half-wave rectifier using 1 diode

The output DC voltage of a half-wave rectifier can be calculated with the following two ideal equations:

$$V_{\text{rms}} = \frac{V_{\text{peak}}}{2} \tag{5-1}$$

$$V_{\text{dc}} = \frac{V_{\text{peak}}}{\pi} \tag{5-2}$$

3. Full-Wave Rectification

A full-wave rectifier converts the whole of the input waveform to one of constant polarity (positive or negative) at its output. Full-wave rectification converts both polarities of the input waveform to DC (direct current), and is more efficient. However, in a circuit with a non-center tapped transformer, four diodes are required instead of the one needed for half-wave rectification (See semiconductors, diode). Four diodes arranged this way are called a diode bridge or bridge rectifier (Fig. 5.2).

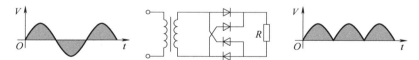

Fig. 5.2　Graetz bridge rectifier: a full-wave rectifier using 4 diodes

For single-phase AC, if the transformer is center-tapped, then two diodes back-to-back (i.e. anodes-to-anode or cathode-to-cathode) can form a full-wave rectifier. Twice as many windings are required on the transformer secondary to obtain the same output voltage compared to the bridge rectifier above (Fig. 5.3).

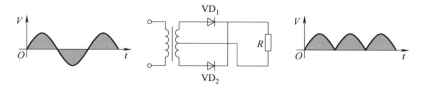

Fig. 5.3　Full-wave rectifier using a center tap transformer and 2 diodes

A very common vacuum tube rectifier configuration contained one cathode and twin anodes inside a single envelope; in this way, the two diodes required only one vacuum tube.

For three-phase AC, six diodes are used. Typically there are three pairs of diodes, each pair, though, is not the same kind of double diode that would be used for a full wave single-phase rectifier. Instead the pairs are in series (anode to cathode). Typically, commercially available double diodes have four terminals so the user can configure them as single-phase split supply use, for half a bridge, or for three-phase use.

Most devices that generate alternating current (such devices are called alternators) generate three-phase AC. For example, an automobile alternator has six diodes inside it to function as a full-wave rectifier for battery charging applications. Fig. 5.4 showed the input three-phase signals and six-pulse output waveform of half & wave rectifier.

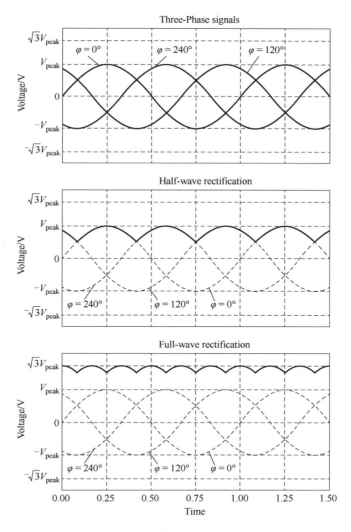

Fig. 5.4 Three-phase AC input, half-& full-wave rectified DC output waveforms

The average and root-mean-square output voltages of an ideal single phase full wave rectifier can be calculated as:

$$V_{dc} = V_{av} = \frac{2V_{peak}}{\pi} \tag{5-3}$$

$$V_{rms} = \frac{V_{peak}}{\sqrt{2}} \tag{5-4}$$

For an ideal three-phase full-wave rectifier (Fig. 5.5), the average output voltage is:

$$V_{dc} = V_{av} = \frac{3\sqrt{3}V_{peak}}{\pi}\cos\alpha \tag{5-5}$$

Where

V_{dc}, V_{av}—the average of DC output voltage, the average of AC output voltage;

V_p—the peak value of half wave;

V_{rms}—the root-mean-square value of output voltage;

$\pi \approx 3.14159$;

α—firing angle of the thyristor.

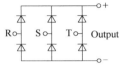

Fig. 5.5 A three-phase bridge rectifier

4. Rectifier Output Smoothing

While half-wave and full-wave rectification suffice to deliver a form of DC output, neither produces constant-voltage DC. In order to produce steady DC from a rectified AC supply, a smoothing circuit or filter is required. In its simplest form this can be just a reservoir capacitor or smoothing capacitor, placed at the DC output of the rectifier (Fig. 5.6). There will still remain an amount of AC ripple voltage where the voltage is not completely smoothed. Sizing of the capacitor represents a tradeoff. For a given load, a larger capacitor will reduce ripple but will cost more and will create higher peak currents in the transformer secondary and in the supply feeding it. In extreme cases where many rectifiers are loaded onto a power distribution circuit, it may prove difficult for the power distribution authority to maintain a correctly shaped sinusoidal voltage curve.

For a given tolerable ripple the required capacitor size is proportional to the load current and inversely proportional to the supply frequency and the number of output peaks of the rectifier per input cycle. The load current and the supply frequency are generally outside the control of the designer of the rectifier system but the number of peaks per input cycle can be affected by the choice of rectifier design.

A half-wave rectifier will only give one peak per cycle and for this and other reasons is only used in very small power supplies. A full wave rectifier achieves two peaks per cycle and this is the best that can be done with single-phase input. For three-phase inputs a three-phase bridge will give

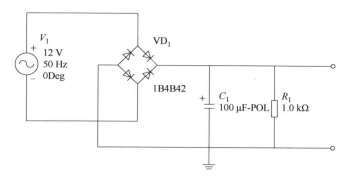

Fig. 5.6 RC-filter rectifier

six peaks per cycle and even higher numbers of peaks can be achieved by using transformer networks placed before the rectifier to convert to a higher phase order.

5. **Configurations Rectifying**

In a common configuration, called monopole, one of the terminals of the rectifier is connected to earth ground (Fig. 5.7). The other terminal, at a potential high above or below ground, is connected to a transmission line. The earthed terminal may be connected to the corresponding connection at the inverting station by means of a second conductor.

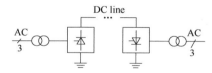

Fig. 5.7 Block diagram of a monopole system with earth return

If no metallic conductor is installed, current flows in the earth between the earth electrodes at the two stations. Therefore it is a type of single wire earth return. The issues surrounding earth-return current include:

1) Electrochemical corrosion of long buried metal objects such as pipelines.

2) Underwater earth-return electrodes in seawater may produce chlorine or otherwise affect water chemistry.

3) An unbalanced current path may result in a net magnetic field, which can affect magnetic navigational compasses for ships passing over an underwater cable.

These effects can be eliminated with installation of a metallic return conductor between the two ends of the monopolar transmission line. Since one terminal of the converters is connected to earth, the return conductor need not be insulated for the full transmission voltage which makes it less costly than the high-voltage conductor. Use of a metallic return conductor is decided based on economic, technical and environmental factors.

Modern monopolar systems for pure overhead lines carry typically 1500 MW. If underground or underwater cables are used, the typical value is 600 MW.

Most monopolar systems are designed for future bipolar expansion. Transmission line towers may be designed to carry two conductors, even if only one is used initially for the monopole transmission system. The second conductor is either unused, used as electrode line or connected in parallel with the other (as in case of Baltic-Cable).

In bipolar transmission, a pair of conductors is used, each at a high potential with respect to ground, in opposite polarity (Fig. 5.8). Since these conductors must be insulated for the full voltage, transmission line cost is higher than a monopole with a return conductor. However, there are a number of advantages to bipolar transmission which can make it the attractive option.

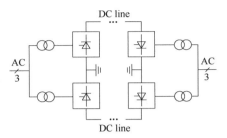

Fig. 5.8 Block diagram of a bipolar system that also has an earth return

1) Under normal load, negligible earth-current flows, as in the case of monopolar transmission with a metallic earth-return. This reduces earth return loss and environmental effects.

2) When a fault develops in a line, with earth return electrodes installed at each end of the line, approximately half the rated power can continue to flow using the earth as a return path, operating in monopolar mode.

3) Since for a given total power rating each conductor of a bipolar line carries only half the current of monopolar lines, the cost of the second conductor is reduced compared to a monopolar line of the same rating.

4) In very adverse terrain, the second conductor may be carried on an independent set of transmission towers, so that some power may continue to be transmitted even if one line is damaged.

5) A bipolar system may also be installed with a metallic earth return conductor. Bipolar systems may carry as much as 3200 MW at voltages of +/-600 kV. Submarine cable installations initially commissioned as a monopole may be upgraded with additional cables and operated as a bipole.

6) A block diagram of a bipolar HVDC transmission system, between two stations designated A and B, is shown in Fig. 5.9. AC represents an alternating current network, CON represents a converter valve, either rectifier or inverter, TR represents a power transformer, DCTL is the direct-current transmission line conductor, DCL is a direct-current filter inductor, BS represents a bypass switch, and PM represents power factor correction and harmonic filter networks required at both ends of the link. The DC transmission line may be very short in a back-to-back link, or extend hundreds of miles (km) overhead, underground or underwater. One conductor of the DC line may be replaced by connections to earth ground.

A bipolar scheme can be implemented so that the polarity of one or both poles can be changed. This allows the operation as two parallel monopoles. If one conductor fails, transmission can still continue at reduced capacity. Losses may increase if ground electrodes and lines are not designed for the extra current in this mode. To reduce losses in this case, intermediate switching

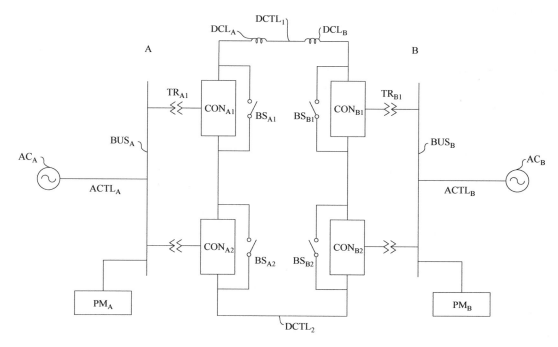

Fig. 5.9 Block diagram of a bipolar HVDC transmission system

stations may be installed, at which line segments can be switched off or parallelized. This was done at Inga-Shaba HVDC.

5.2.3 Inverter in HVDC

An inverter is an electrical device that converts DC to AC; the converted AC can be at any required voltage and frequency with the use of appropriate transformers, switching, and control circuits.

Solid-state inverters have no moving parts and are used in a wide range of applications, from small switching power supplies in computers, to large electric utility high-voltage direct current applications that transport bulk power. **Inverters are commonly used to supply AC power from DC sources such as solar panels or batteries.**

The inverter performs the opposite function of a rectifier.

1. Types

(1) Modified Sine Wave

The output of a modified sine wave inverter is similar to a square wave output except that the output goes to zero volts for a time before switching positive or negative. It is simple and low cost (~0.10USD/W) and is compatible with most electronic devices, except for sensitive or specialized equipment, such as certain laser printers, fluorescent lighting, audio equipment.

Most AC motors will run off this power source albeit at a reduction in efficiency of approximately 20%.

(2) Pure Sine Wave

A pure sine wave inverter produces a nearly perfect sine wave output (<3% total harmonic distortion) that is essentially the same as utility-supplied grid power. Thus it is compatible with all AC electronic devices. This is the type used in grid-tie inverters. Its design is more complex, and costs more per unit power. The electrical inverter is a high-power electronic oscillator. It is so named because early mechanical AC to DC converters were made to work in reverse, and thus were "inverted", to convert DC to AC.

(3) Grid Tie Inverter

A grid tie inverter is a sine wave inverter designed to inject electricity into the electric power distribution system. Such inverters must synchronize with the frequency of the grid. They usually contain one or more maximum power point tracking features to extract the maximum amount of power, and also include safety features.

2. Applications

(1) DC Power Source Utilization

Inverter designed to provide 115 V AC from the 12 V DC source provided in an automobile. The unit shown provides up to 1.2 amperes of alternating current, or enough to power two sixty watt light bulbs.

An inverter converts the DC electricity from sources such as batteries, solar panels, or fuel cells to AC electricity. The electricity can be at any required voltage; in particular it can operate AC equipment designed for mains operation, or rectified to produce DC at any desired voltage. Micro-inverters convert direct current from individual solar panels into alternating current for the electric grid. They are grid tie designs by default.

(2) Uninterruptible Power Supplies

An uninterruptible power supply (UPS) uses batteries and an inverter to supply AC power when main power is not available. When main power is restored, a rectifier supplies DC power to recharge the batteries.

(3) Induction Heating

Inverters convert low frequency main AC power to higher frequency for use in induction heating. To do this, AC power is first rectified to provide DC power. The inverter then changes the DC power to high frequency AC power.

(4) HVDC Power Transmission

With HVDC power transmission, AC power is rectified and high voltage DC power is transmitted to another location. At the receiving location, an inverter in a static inverter plant converts the power back to AC.

(5) Variable-Frequency Drives

A variable-frequency drive controls the operating speed of an AC motor by controlling the frequency and voltage of the power supplied to the motor. An inverter provides the controlled power. In most cases, the variable-frequency drive includes a rectifier so that DC power for the inverter can be provided from main AC power. Since an inverter is the key component, variable-frequency drives are sometimes called inverter drives or just inverters.

(6) Electric Vehicle Drives

Adjustable speed motor control inverters are currently used to power the traction motors in some electric and diesel-electric rail vehicles as well as some battery electric vehicles and hybrid electric highway vehicles such as the Toyota Prius and Fisker Karma. Various improvements in inverter technology are being developed specifically for electric vehicle applications. In vehicles with regenerative braking, the inverter also takes power from the motor (now acting as a generator) and stores it in the batteries.

A transformer allows AC power to be converted to any desired voltage, but at the same frequency. Inverters, plus rectifiers for DC, can be designed to convert from any voltage, AC or DC, to any other voltage, also AC or DC, at any desired frequency. The output power can never exceed the input power, but efficiencies can be high, with a small proportion of the power dissipated as waste heat.

3. Circuit Description

(1) Basic Designs

In one simple inverter circuit, DC power is connected to a transformer through the centre tap of the primary winding. A switch is rapidly switched back and forth to allow current to flow back to the DC source following two alternate paths through one end of the primary winding and then the other. The alternation of the direction of current in the primary winding of the transformer produces alternating current (AC) in the secondary circuit.

The electromechanical version of the switching device includes two stationary contacts and a spring supported moving contact. The spring holds the movable contact against one of the stationary contacts and an electromagnet pulls the movable contact to the opposite stationary contact. The current in the electromagnet is interrupted by the action of the switch so that the switch continually switches rapidly back and forth. This type of electromechanical inverter switch, called a vibrator or buzzer, was once used in vacuum tube automobile radios. A similar mechanism has been used in door bells, buzzers and tattoo guns.

As they became available with adequate power ratings, transistors and various other types of semiconductor switches have been incorporated into inverter circuit designs.

(2) Output Waveforms

The switch in the simple inverter described above, when not coupled to an output transformer, produces a square voltage waveform due to its simple off and on nature as opposed to the sinusoidal waveform that is the usual waveform of an AC power supply. Using Fourier analysis, periodic waveforms are represented as the sum of an infinite series of sine waves (Fig. 5.10). The sine wave that has the same frequency as the original waveform is called the fundamental component. The other sine waves, called harmonics, which are included in the series have frequencies that are integral multiples of the fundamental frequency.

The quality of the inverter output waveform can be expressed by using the Fourier analysis data to calculate the total harmonic distortion (THD). THD is the square root of the sum of the squares

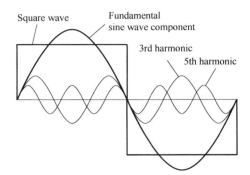

Fig. 5.10　Square waveform with fundamental sine wave component, 3rd harmonic and 5th harmonic

of the harmonic voltages divided by the fundamental voltage:

$$\mathrm{THD} = \frac{\sqrt{V_2^2 + V_3^2 + V_4^2 + \cdots + V_n^2}}{V_1} \quad (5\text{-}6)$$

The quality of output waveform that is needed from an inverter depends on the characteristics of the connected load. Some loads need a nearly perfect sine wave voltage supply in order to work properly. Other loads may work quite well with a square wave voltage.

H-bridge inverter circuit with transistor switches and antiparallel diodes is shown in Fig. 5.11. There are many different power circuit topologies and control strategies used in inverter designs. Different design approaches address various issues that may be more or less important depending on the way that the inverter is intended to be used.

(3) **Advanced Designs**

The issue of waveform quality can be addressed in many ways. Capacitors and inductors can be used to filter the waveform. If the design includes a transformer, filtering can be applied to the primary or the secondary side of the transformer or to both sides. Low-pass filters are applied to allow the fundamental component of the waveform to pass to the output while limiting the passage of the harmonic components. If the inverter is designed to

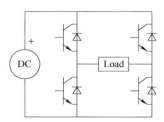

Fig. 5.11　H-bridge inverter circuit with transistor switches and antiparallel diodes

provide power at a fixed frequency, a resonant filter can be used. For an adjustable frequency inverter, the filter must be tuned to a frequency that is above the maximum fundamental frequency.

Since most loads contain inductance, feedback rectifiers or antiparallel diodes are often connected across each semiconductor switch to provide a path for the peak inductive load current when the switch is turned off. The antiparallel diodes are somewhat similar to the freewheeling diodes used in AC/DC converter circuits.

4. **Three-Phase Inverters**

Three-phase inverters are used for variable-frequency drive applications and for high power applications such as HVDC power transmission. A basic three-phase inverter consists of three single-phase inverter switches each connected to one of the three load terminals (Fig. 5.12). For the most basic control scheme, the operation of the three switches is coordinated so that one switch operates at each 60° point of the fundamental output waveform. This creates a line-to-line output waveform that has six steps. The six-step waveform has a zero-voltage step between the positive and negative sections of the square-wave such that the harmonics that are multiples of three are

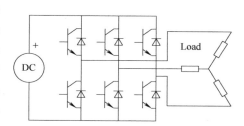

Fig. 5.12 Three-phase inverter with wye connected load

eliminated as described above. When carrier-based PWM techniques are applied to six-step waveforms, the basic overall shape, or envelope, of the waveform is retained so that the 3rd harmonic and its multiples are cancelled.

To construct inverters with higher power ratings, two six-step three-phase inverters can be connected in parallel for a higher current rating or in series for a higher voltage rating. In either case, the output waveforms are phase shifted to obtain a 12-step waveform. If additional inverters are combined, an 18-step inverter is obtained with three inverters etc. Although inverters are usually combined for the purpose of achieving increased voltage or current ratings, the quality of the waveform is improved as well.

5. Controlled Rectifier Inverters

Since early transistors were not available with sufficient voltage and current ratings for most inverter applications, it was the 1957 introduction of the thyristor or SCR that initiated the transition to solid state inverter circuits.

The commutation requirements of SCRs are a key consideration in SCR circuit designs. SCRs do not turn off or commutate automatically when the gate control signal is shut off. They only turn off when the forward current is reduced to below the minimum holding current, which varies with each kind of SCR, through some external process. For SCRs connected to an AC power source, commutation occurs naturally every time the polarity of the source voltage reverses. SCRs connected to a DC power source usually require a means of forced commutation that forces the current to zero when commutation is required. The least complicated SCR circuits employ natural commutation rather than forced commutation. With the addition of forced commutation circuits, SCRs have been used in the types of inverter circuits described in Fig. 5.13.

In applications where inverters transfer power from a DC power source to an AC power source, it is possible to use AC-to-DC controlled rectifier circuits operating in the inversion mode. In the inversion mode, a controlled rectifier circuit operates as a line commutated inverter. This type of operation can be used in HVDC power transmission systems and in regenerative braking operation of motor control systems.

A back-to-back station (or B2B for short) is a plant in which both static inverters and rectifiers are in the same area, usually in the same building. The length of the direct current line is kept as short as possible. HVDC back-to-back stations are used for:

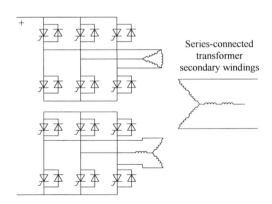

Fig. 5.13 12-pulse line-commutated inverter circuit

1) Coupling of electricity mains of different frequency [as in Japan; and the GCC interconnection between UAE (50 Hz) and Saudi Arabia (60 Hz) under construction in 2009-2011].

2) Coupling two networks of the same nominal frequency but no fixed phase relationship (as until 1995/96 in Etzenricht, Dürnrohr, Vienna, and the Vyborg HVDC scheme).

3) Different frequency and phase number (as a replacement for traction current converter plants).

The DC voltage in the intermediate circuit can be selected freely at HVDC back-to-back stations because of the short conductor length. The DC voltage is as low as possible, in order to build a small valve hall and to avoid series connections of valves. For this reason at HVDC back-to-back stations valves with the highest available current rating are used.

5.3 Application of HVDC

The most common configuration of an HVDC link is two inverter/rectifier stations connected by an overhead power line. This is also a configuration commonly used in connecting unsynchronised grids, in long-haul power transmission, and in undersea cables.

Multi-terminal HVDC links, connecting more than two points, are rare. The configuration of multiple terminals can be series, parallel, or hybrid (a mixture of series and parallel). Parallel configuration tends to be used for large capacity stations, and series for lower capacity stations. An example is the 2000 MW Quebec-New England transmission system opened in 1992, which is currently the largest multi-terminal HVDC system in the world.

A scheme patented in 2004 (current modulation of direct current transmission lines) is intended for conversion of existing AC transmission lines to HVDC. Two of the three circuit conductors are operated as a bipole. The third conductor is used as a parallel monopole, equipped with reversing valves (or parallel valves connected in reverse polarity). The parallel monopole periodically relieves current from one pole or the other, switching polarity over a span of several minutes. The bipole conductors would be loaded to either 1.37 or 0.37 of their thermal limit, with the parallel monopole always carrying ±1 times its thermal limit current. The combined RMS heating effect is as if each of the conductors is always carrying 1.0 of its rated current. This allows heavier currents to be carried by the bipole conductors, and full use of the installed third conductor for energy transmission. High currents can be circulated through the line conductors even when load demand is low, for removal of ice.

As of 2005, no tri-pole conversions are in operation, although a transmission line in India has been converted to bipole HVDC. Cross-Skagerrak consists of 3 poles, from which 2 are switched in parallel and the third uses an opposite polarity with a higher transmission voltage. A similar arrangement is HVDC Inter-Island, but it consists of 2 parallel-switched inverters feeding in the same pole and a third one with opposite polarity and higher operation voltage.

5.3.1 Corona Discharge

Corona discharge is the creation of ions in a fluid (such as air) by the presence of a strong electric field. Electrons are torn from neutral air, and either the positive ions or the electrons are attracted to the conductor, while the charged particles drift. This effect can cause considerable power loss, create audible and radio-frequency interference, generate toxic compounds such as oxides of nitrogen and ozone, and bring forth arcing.

Both AC and DC transmission lines can generate coronas, in the former case in the form of oscillating particles, in the latter a constant wind. Due to the space charge formed around the conductors, an HVDC system may have about half the loss per unit length of a high voltage AC system carrying the same amount of power. With monopolar transmission the choice of polarity of the energized conductor leads to a degree of control over the corona discharge. In particular, the polarity of the ions emitted can be controlled, which may have an environmental impact on particulate condensation (particles of different polarities have a different mean-free path). Negative coronas generate considerably more ozone than positive coronas, and generate it further downwind of the power line, creating the potential for health effects. The use of a positive voltage will reduce the ozone impacts of monopole HVDC power lines.

The controllability of current-flow through HVDC rectifiers and inverters, their application in connecting unsynchronized networks, and their applications in efficient submarine cables mean that HVDC cables are often used at national boundaries for the exchange of power (in North America, HVDC connections divide much of Canada and the United States into several electrical regions that cross national borders, although the purpose of these connections is still to connect unsynchronized AC grids to each other). Offshore windfarms also require undersea cables, and their turbines are unsynchronized. In very long-distance connections between just two points, for example around the remote communities of Siberia, Canada, and the Scandinavian North, the decreased line-costs of HVDC also makes it the usual choice.

5.3.2 AC Network Interconnections

AC transmission lines can interconnect only synchronized AC networks that oscillate at the same frequency and in phase. Many areas that wish to share power have unsynchronized networks. The power grids of the UK, Northern Europe and continental Europe are not united into a single synchronized network. Japan has 50 Hz and 60 Hz networks. Continental North America, while operating at 60 Hz throughout, is divided into regions which are unsynchronized: East, West, Texas, Quebec, and Alaska. Brazil and Paraguay, which share the enormous Itaipu Dam hydroelectric plant, operate on 60 Hz and 50 Hz respectively. However, HVDC systems make it possible to interconnect unsynchronized AC networks, and also add the possibility of controlling AC voltage and reactive power flow.

A generator connected to a long AC transmission line may become unstable and fall out of synchronization with a distant AC power system. An HVDC transmission link may make it economically feasible to use remote generation sites. Wind farms located off-shore may use HVDC systems to collect power from multiple unsynchronized generators for transmission to the shore by an underwater cable.

In general, however, an HVDC power line will interconnect two AC regions of the power-distribution grid. Machinery to convert between AC and DC power adds a considerable cost in power transmission. The conversion from AC to DC is known as rectification, and from DC to AC as inversion. Above a certain break-even distance (about 50 km for submarine cables, and perhaps 600-800 km for overhead cables), the lower cost of the HVDC electrical conductors outweighs the cost of the electronics. The conversion electronics also present an opportunity to effectively manage the power grid by means of controlling the magnitude and direction of power flow. An additional advantage of the existence of HVDC links, therefore, is potential increased stability in the transmission grid. Two HVDC lines cross near Wing, North Dakota are shown in Fig. 5.14.

5.3.3 Renewable Electricity Superhighways

A number of studies have highlighted the potential benefits of very wide area super grids based on HVDC since they can mitigate the effects of intermittency by averaging and smoothing the outputs of large numbers of geographically dispersed wind farms or solar farms. Czisch's study concludes that a grid covering the fringes of Europe could bring 100% renewable power (70% wind, 30% biomass) at close to today's prices. There has been debate over the technical feasibility of this proposal and the political risks involved in energy transmission across a large number of international borders.

The construction of such green power superhighways is advocated in a white paper that was released by the American Wind Energy Association (AWEA) and the Solar Energy Industries Association (SEIA). In January 2009, the European Commission proposed €300 million to subsidize the development of HVDC links between Ireland, Britain, the Netherlands, Germany, Denmark, and Sweden, as part of a wider €1.2 billion package supporting links to off-shore wind farms and cross-border interconnectors throughout Europe. Meanwhile the recently founded Union of the Mediterranean has embraced a Mediterranean Solar Plan to import large amounts of concentrating solar power into Europe from North Africa and the Middle East.

Fig. 5.14   Two HVDC lines cross near Wing, North Dakota

5.3.4   Voltage Sourced Converters (VSC)

The development of insulated gate bipolar transistors (IGBT) and gate turn-off thyristors (GTO) has made smaller HVDC systems economical. These may be installed in existing AC grids for their role in stabilizing power flow without the additional short-circuit current that would be produced by an additional AC transmission line. The manufacturer ABB calls this concept "HVDC Light", while Siemens calls a similar concept "HVDC PLUS (power link universal system)". They have extended the use of HVDC down to blocks as small as a few tens of megawatts and lines as short as a few score kilometres of overhead line. There are several different variants of voltage-sourced converter (VSC) technology: most "HVDC Light" installations use pulse width modulation but the most recent installations, along with "HVDC PLUS", are based on multilevel switching. The latter is a promising concept as it allows reducing the filtering efforts to a minimum. At the moment, the line filters of typical converter stations cover nearly half of the area of the whole station.

5.4   HVDC in China

A HVDC electric power transmission system uses direct current for the bulk transmission of electrical power, in contrast with the more common alternating current systems. For long-distance transmission, HVDC systems may be less expensive and suffer lower electrical losses. For underwater power cables, HVDC avoids the heavy currents required by the cable capacitance. For shorter distances, the higher cost of DC conversion equipment compared to an AC system may still be warranted, due to other benefits of direct current links. HVDC allows power transmission between unsynchronized AC distribution systems, and can increase system stability by preventing cascading failures from propagating from one part of a wider power transmission grid to another.

The modern form of HVDC transmission uses technology developed extensively in the 1930s in Sweden (ASEA). Early commercial installations included one in the Soviet Union in 1951 between Moscow and Kashira, and a 10-20 MW system between Gotland and mainland Sweden in 1954. The longest HVDC link in the world is currently the Xiangjiaba-Shanghai 1907 km 6400 MW link connecting the Xiangjiaba Dam to Shanghai, in China. In 2012, the longest HVDC link will be the Rio Madeira link connecting the Amazonas to the São Paulo area where the length of the DC line is over 2500 km.

High voltage (in either DC or AC electrical power transmission applications) is used for electric power transmission to reduce the energy lost in the resistance of the wires. High voltages cannot easily be used for lighting and motors, and so transmission-level voltages must be reduced to values compatible with end-use equipment. Transformers are used to change the voltage level in AC transmission circuits. The competition between the DC of Thomas Edison and the AC of Nikola Tesla and George Westinghouse was known as the War of Currents, with AC becoming dominant.

Practical manipulation of high power HVDC became possible with the development of high power electronic rectifier devices such as mercury arc valves and, more recently starting in the 1970s, high power semiconductor devices such as high power thyristors and 21st century high power variants such as integrated gate-commutated thyristors (IGCT), MOS controlled thyristors (MCT) and gate turn-off thyristors (GTO). A similar high power transistor device called the insulated-gate bipolar transistors has recently been used in these applications.

VOCABULARY

1. HVDC　高压直流输电
2. solid-state power device　固态功率设备
3. converter　*n.* 变流器
4. anode　*n.* 阳极
5. cathode　*n.* 阴极
6. forward current　正向导通电流
7. diode　*n.* 二极管
8. P-N junction　PN 结
9. silicon-controlled rectifier　晶闸管（整流器）
10. leakage current　漏电流
11. emitter　*n.* 发射极
12. collector　*n.* 集电极
13. rectifier　*n.* 整流器
14. full-wave　全波
15. harmonic　*n.* 谐波
16. vacuum tube diode　真空二极管
17. selenium rectifier　硒整流器
18. half-wave　半波
19. transformer　*n.* 变压器
20. center-tapped　中间抽头的
21. solar panel　太阳能板
22. sine wave　正弦波
23. grid tie inverter　并网逆变器
24. electric grid　电网
25. uninterruptible power supply (UPS)　不间断电源
26. inverter　*n.* 逆变器
27. electromechanical　*adj.* 机电的

28. total harmonic distortion（THD） 总谐波畸变率
29. square wave 方波
30. H-bridge inverter H 桥逆变器
31. resonant *adj.* 谐振的
32. antiparallel diode 反并联二极管
33. freewheeling diode 续流二极管
34. pulse-width modulation（PWM） 脉宽调制
35. semiconductor *n.* 半导体
36. alternator *n.* 交流发电机
37. root-mean-square 均方根
38. ripple *n.* 纹波
39. impedance *n.* 阻抗
40. fluorescent lighting 荧光灯照明
41. maximum power point tracking 最大功率点跟踪
42. micro-inverter 微型逆变器
43. static inverter plant 静态逆变站
44. variable-frequency drive 变频驱动
45. vibrator *n.* 振荡器
46. buzzer *n.* 蜂鸣器
47. sinusoidal *adj.* 正弦的
48. direct current 直流
49. alternating current 交流
50. underwater power cable 海底电力电缆
51. mercury arc 汞弧
52. integrated gate-commutated thyristor 集成门极换流晶闸管
53. MOS controlled thyristor MOS 可控晶闸管
54. gate turn-off thyristor 门极可关断晶闸管
55. insulated-gate bipolar transistor 绝缘栅双极型晶体管

NOTES

1. Since the power lost as heat in the wires is proportional to the square of the current, but does not depend in any major way on the voltage delivered by the power line, doubling the voltage in a power system reduces the line-loss loss per unit of electrical power delivered by a factor of 4.

因为线路中的热功率损耗与电流的二次方成正比，但是与电线的电压没有任何关系。因此如果将电力系统中的电压变为原来的两倍可以相应地使线路中的损耗减少为原来的 1/4。

2. High voltages cannot easily be used for lighting and motors, and so transmission-level voltages must be reduced to values compatible with end-use equipment.

高压不能轻易地用在照明和电机上，因此传输的电压必须降低到终端设备能接受的电压值。

3. However, when direct current is used, the cable capacitance is charged only when the cable is first energized or when the voltage is changed; there is no steady-state additional current required.

然而当采用直流时，电缆电容只有当电缆刚开始通电或者电压改变时才被充电，不需要其他稳态的附加电流。

4. Changes in load that would cause portions of an AC network to become unsynchronized and separately would not similarly affect a DC link, and the power flow through the DC link would tend to stabilize the AC network.

负载的变化使得交流电网中某些部分变得不同步，但相应地不会同样影响直流网络，并且流过直流网络的潮流能使交流电网稳定（相对于交流潮流）。

5. When controlled rectifier circuits are operated in the inversion mode, they would be classified by pulse number also. Rectifier circuits that have a higher pulse number have reduced harmonic content in the AC input current and reduced ripple in the DC output voltage.

当可控整流器电路工作在逆变状态时,电路就会按照脉动数分类。高脉动数的整流器电路减少了交流输入电流的谐波含量,降低了直流输出电压的纹波。

6. For a given tolerable ripple the required capacitor size is proportional to the load current and inversely proportional to the supply frequency and the number of output peaks of the rectifier per input cycle.

在给定容许纹波水平条件下,所需电容容值与负载电流成正比,与电源频率以及每个输入周期整流器输出电压波头数成反比。

7. Since one terminal of the converters is connected to earth, the return conductor need not be insulated for the full transmission voltage which makes it less costly than the high-voltage conductor.

因为变流器的一端接地,回线不需要对整个传输电压绝缘,这使得其造价比高压导线低。

8. Solid-state inverters have no moving parts and are used in a wide range of applications, from small switching power supplies in computers, to large electric utility high-voltage direct current applications that transport bulk power.

固态逆变器没有机械运动的部分,并且应用广泛。从计算机里的小型开关电源到传输大功率电能的大型高压直流输电方面都有应用。

9. An inverter provides the controlled power. In most cases, the variable-frequency drive includes a rectifier so that DC power for the inverter can be provided by main AC power.

逆变器提供可控的电能。在大多数情况下,变频驱动包含有一个整流器,这样逆变器需要的直流电能可由交流主电源来提供。

10. If the design includes a transformer, filtering can be applied to the primary or the secondary side of the transformer or to both sides. Low-pass filters are applied to allow the fundamental component of the waveform to pass to the output while limiting the passage of the harmonic components.

如果设计中含有变压器,则可以在变压器的一次侧或者二次侧或者两侧都加装滤波器。低通滤波器在输出过程中允许波形中的基波成分通过,同时限制谐波成分通过。

11. The commutation requirements of SCRs are a key consideration in SCR circuit designs. SCRs do not turn off or commutate automatically when the gate control signal is shut off.

在SCR电路设计中,换相条件应该是主要考虑的方面。SCR在门极控制信号关断时不会自动关断或者换相。

12. The development of insulated gate bipolar transistors (IGBT) and gate turn-off thyristors (GTO) has made smaller HVDC systems economical. These may be installed in existing AC grids for their role in stabilizing power flow without the additional short-circuit current that would be produced by an additional AC transmission line.

IGBT和GTO的发展使得高压直流系统更为经济。这些器件可以安装在现有的交流电网中起稳定潮流的作用,不会产生附加的短路电流,而新增交流传输线则不可避免存在短路电流。

Chapter 6
Flexible AC Transmission (FACT) Systems

6.1 Introduction

Tighter control of power flow and increased use of transmission capacity are key benefits of new thyristor-based controllers. A power line can function nearer its top thermal rating if regulated by flexible AC transmission system controllers.

The electric utilities' systems for transmitting and distributing power are entering a period of change. Their operation is fine-tuned to an unprecedented degree, by the application of power electronics, microprocessors and microelectronics in general, and communications. These technologies will make the transmission and distribution of electricity more reliable, more controllable, and more efficient.

Acting for the US electric utility industry, the Electric Power Research Institute (EPRI) in Palo Alto, Calif., has led the way in this area with several thrusts, not the least being the flexible AC transmission system, known in the industry as FACTS.

Power electronic based flexible AC transmission system (FACTS) devices provide proven solutions to meet these challenges. When used in combination with wide area control systems (WACS) it is possible to increase performance benefits further. FACTS controllers can be classified based on the power electronic technology used. Existing installations that use line-commutated thyristor based technologies are either static var compensators (SVC) or thyristor controlled series compensators (TCSC). FACTS devices can also be based on voltage source converters, which utilize self-commutated devices such as gate turn off thyristors (GTO), gate commutated thyristors (GCT), or insulated gate bipolar transistors (IGBT). Existing VSC-based FACTS installations using can be categorized as static synchronous compensators (STATCOM), unified power flow controllers (UPFC), and convertible static compensators (CSC). In some STATCOM behavior is included as a secondary control option in devices such as power electronic interfaces for wind turbines (such as the field control on a DFIG) or the power electronic interface for a energy storage system such as a battery energy storage system (BESS), a superconducting magnetic energy storage system (SMES) or a flywheel energy storage system. Voltage source converter based high voltage direct current (VSC-HVDC) transmission also exhibits dynamic reactive control capabilities which are utilized as well.

The development of flexible transmission system is akin to high-voltage DC and related thyristor developments, designed to overcome the limitations of the present mechanically controlled AC power transmission systems. By using reliable, high-speed power electronic controllers, the technology offers utilities five opportunities to increased efficiency:

1) Greater control of power, so that it flows on the prescribed transmission routes.

2) Secure loading (but not overloading) of transmission lines to levels nearer their thermal limits.

3) Greater ability to transfer power between controlled areas, so that the generation reserve margin, typically 18%, may be reduced to 15% or less.

4) Prevention of cascading outages by limiting the effects of faults and equipment failure.

5) Damping of power system oscillations, which could damage equipment and/or limit usable transmission capacity.

Advantages and savings must be weighed against the cost of the power electronic controllers required. At about US $50-$100 per kilovolt-ampere (kV·A) rating of the thyristor-based controllers, the typical capital cost of these controllers can already be afforded for some utility applications (roughly speaking, the cost per kilovolt-ampere decreases with an increase in the size of the controller).

The flexible system owes its tighter transmission control to its ability to manage the interrelated parameters that constrain today's systems, including series impedance, shunt impedance, phase angle, and the occurrence of oscillations at various frequencies below the rated frequency. By adding to flexibility in this way, the controllers enable a transmission line to function nearer its thermal rating. For example, a 500 kV line may have a loading limit of 1000-2000 MW for safe operation, but a thermal limit of 3000 MW.

It is often not possible both to overcome these constraints and to maintain the required system reliability by conventional mechanical means alone, such as tap changers, phase shifters, and switched capacitors and reactors (inductors). Granted, mechanical controllers are on the whole less expensive, but they increasingly need to be supplemented by rapidly responding power electronics controllers.

The new technology is not a single, high power electronic controller, but rather a collection of controllers, which can be applied individually or collectively in a specific power system to control the five interrelated functions already mentioned. The thyristor is their basic element, just as the transistor is the basic element for a whole variety of microelectronic circuits. Because all controllers for the flexible transmission system are applications of similar technology, their use will eventually benefit from volume production and further development of high-power electronics.

Electric power networks integrate generation and load centers within each utility system and share power with vast regional grids through interconnections among neighboring systems. The purpose of this is to take advantage of the diversity of loads, changes in peak demand due to weather and time differences, the availability of different generation reserves in various geographic regions, power-sharing arrangements among utilities, shifts in fuel prices, regulatory changes, and other discrepancies.

1. Transmission Links

By facilitating bulk power transfers, these interconnected networks help minimize the need to enlarge power plants and enable neighboring utilities and regions to buy and sell power among themselves. Thus, the electric power transmission network is essential for reliable, low-cost power. Conversely, inadequate transmission will result in less reliable, more costly.

The demands placed on the transmission network have grown in recent years, and will go on growing, both because non-utility generators (NUG) are entering the market in increasing numbers and because competition among the utilities themselves has heightened. Making matters worse is the extreme difficulty of acquiring new rights-of-way. Although the flexible transmission technology can alleviate some of these pressures, it must be stressed that for much capacity expansion, building or upgrading of lines without resorting to flexible transmission technology will still be the most economical way to go. What is of most interest to the transmission planner is the new options opened up by the technology for controlling power and enhancing the usable capacity of present lines through voltage and current upgrading, impedance modification, and phase angle regulation.

2. Free Flow of Power

Many transmission facilities confront one or more limiting network parameters plus the inability to direct power flow at will.

A well-known formula states that the power flow between two points along a transmission line is equal to the product of the voltages at these points, times the sine of the difference between their phase angles, all divided by the transmission line's reactance between the two points. To understand the free flow of power, consider an elementary case in which two generators are sending power to a load center through a network consisting of three lines in a meshed connection (Fig. 6.1a). The lines AB, BC, and AC have continuous ratings of 1000 MW, 1250 MW, and 2000 MW respectively with emergency ratings of twice those figures. If one generator is rated at 2000 MW and the other at 1000 MW, a total of 3000 MW would be delivered to the load center. For the impedances shown, the three lines should carry 600 MW, 1600 MW, and 1400 MW, respectively. Such a situation would overload one line.

Power, in short, flows in accordance with transmission line impedances that bear no direct relationship to transmission ownership, contracts, or thermal limits. The difference between the free-flow path and the contract path is called "loop flow," and is usually characterized by a circulation of power that leaves the available capacity underutilized.

If, however, a capacitor whose reactance is -5Ω at the synchronous frequency is inserted in one line (Fig. 6.1b), it reduces the line's impedance from 10 Ω to 5 Ω, so that power flow through the lines will be 250 MW, 1250 MW, and 1750 MW respectively. It is clear that if the series capacitor is adjustable, then other power flow levels may be realized in accordance with the ownership, contract, and thermal limitations. This capacitor could be modular and mechanically switched. but the number of operations would be severely limited by weir on the mechanical components.

Defining terms

Phase angle regulator: a controller for shifting the phase angle of an ac voltage.

Static condenser: a solid-state controller, functionally equivalent to a synchronous condenser, which can supply capacitive, as well as inductive, reactive power and may have a small stored energy capability for damping oscillations.

Static var compensator: a solid-state controller that regulates reactive (voltampere) power along a transmission line by switching various combinations of capacitors and inductors in parallel with the line.

SSR damper (in power transmission): a solid-state controller that damps subsynchronous resonance oscillations.

Subsynchronous resonance (SSR): one occurring typically around 15-30 Hz, when a natural frequency of a generator's mechanical shaft coincides with an electrical resonance frequency of the transmission network.

Thermal limit (of a transmission line): the line's maximum current-carrying capacity.

Thyristor value: a stack of interconnected thyristors used for switching. (Note: the term originates from high-voltage dc converter applications.)

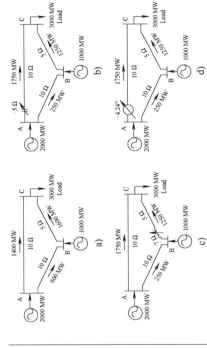

A simple meshed network illustrates hypothetical power flow situations. Here generators at A and B feed a 3000MW load at C via transmission lines AB, with continuous power rating of 1000 MW, BC (1250MW) and AC (2000MW). At the impedances shown, BC will be overloaded, carrying 1600MW(a). Series three-phase thyristor-controlled capacitors in line AC (b), reactors (inductors) in line BC (c), or phase shifters in line AC (d) can restore the flow in BC to no more than its rated 1250 MW.

Fig. 6.1 Free flow of power in simple structure grid

Other complications may arise. A series capacitor in a line may lead to sub-synchronous resonance, typically at 15-30 Hz. This resonance occurs when the mechanical resonance frequency of the shaft of the generator coincides with 60 Hz minus the electrical resonance frequency of the capacitor in series with the total system impedance. If such resonance persists, it soon damages the shaft. Furthermore, while the outage of one line is forcing other lines to operate at their emergency ratings and carry higher loads, power flow oscillations at low frequency (typically 1-2 Hz) may cause generators to lose synchronism, perhaps prompting the system's collapse.

If the series capacitor is thyristor controlled, it can be operated as often as required and can be modulated so rapidly to damp any sub-synchronous resonance conditions, as well as low-frequency oscillations in power flow, and allow the transmission system to go from one steady state condition to another without damage to a generator shaft or the collapse of the system. In other words, a thyristor-controlled series capacitor can greatly enhance the stability of the network. More often than not, though, it is most practical for part of the series compensation to be mechanically controlled and part thyristor controlled, so as to counter the system contraints at least cost.

Similar results may be obtained by increasing the impedance of one of the lines in the same meshed configuration by inserting a 7 Ω reactor (inductor) in series with the line (Fig. 6.1c). Again, a series inductor that is partly mechanically and partly thyristor controlled could serve to adjust the steady-state power flows as well as damp unwanted oscillations.

In either case, a thyristor-controlled phase angle regulator could be installed instead of a series capacitor or a series reactor in any of the three lines to serve the same purpose. Note that neither the inductor nor the phase angle regulator contributes to sub-synchronous resonance. In Fig. 6.1d, the regulator is installed in the third line to reduce the total phase angle difference along the line from 8.5° to 4.2°. As before, a combination of mechanical and thyristor control in the phase angle regulator may minimize cost.

Several controllers are presently being evaluated for flexible transmission systems, while others have been conceptualized but not yet developed. What might be called the first generation of controllers includes two thyristor-based systems that have been used in some utility systems for several years.

3. Static Var Compensator

The first, a static var compensator (SVC) has been used since the 1970s. It addresses the problem of keeping steady-state and dynamic voltages within bounds, and has some ability to control stability, but none to control active power flow. The SVC uses thyristor valves to add or remove shunt-connected reactors and/or capacitors rapidly, often in coordination with mechanically controlled reactors and/or capacitors.

The first application of an SVC to voltage control was demonstrated on the Tri-State G&T System in 1977 by General Electric Co. (GE), which is headquartered in Fairfield, Conn. Another SVC with voltage and stability control, developed with EPRI funding by Westinghouse Electric Corp. of Pittsburgh began operation in 1978 on the Minnesota Power and Light System (Fig. 6.2, top) (incidentally, since EPRI's launching of the flexible AC transmission system strategy in 1986, the market for SVCs has increased substantially.)

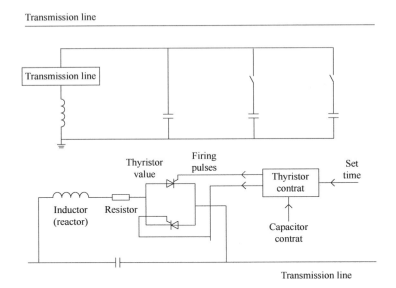

Fig. 6.2 Resonance damper and its application

The second controller in actual use is the NGH-SSR damper, inverted to counter sub-synchronous resonance (SSR). SSR instabilities are at times an undesirable side effect of using a mechanically controlled series capacitor to add up to 80% compensation to a transmission line, the goal being to lower the line's impedance, increase power flow, and expand stability limits. In the early 1970s, after the shaft of a turbine generator belonging to Southern California Edison Co. was damaged by sub-synchronous resonance, the series compensation level on a major 500 kV transmission corridor had to be reduced, so that less power could be transferred over it. Since then, various solutions to the problems of sensing sub-synchronous resonance, emergency-switching, blocking it, and so on, have been adopted.

4. **Resonance Damper**

The NGH-SSR damper consists of a thyristor AC switch (back-to-back thyristors) connected in series with a small inductor and resistor across the series capacitor. The operation of the damper is based on two principles. One is to fire the switch 8.33 ms after each zero of the capacitor's voltage, or half a cycle (180°) at 60 Hz. But if the voltage wave contains other frequencies, some half cycles will be longer than 8.33 ms. In this case the valve firing at 8.33 ms causes some current to flow during the extended part of the half cycle and damps the oscillations. The second principle is to fire the switch somewhat earlier than 8.33 ms, or less than 180° following the zero. Earlier firing causes the impedance of the combined circuit to be more negative than that with the capacitor alone, thus detuning the electric circuit. Furthermore, by the modulation of the firing angle, the impedance can have a powerful damping effect at any unwanted frequency below the main frequency.

In an SSR damper installed in Southern California Edison's Lugo substation (Fig. 6.2, bottom), the thyristors have a modest current rating (15% of load) for continuously by passing the capacitor's wave for only up to the last 10° or so of the voltage wave. With more cooling of the thyristors and appropriate sizing of the reactor or resistor in series with the thyristor switch, this damper forms the basis for the fully rated thyristor-controlled series capacitor described later.

With the recognition of flexible AC transmission technology as a highly effective means of enhancing power systems, a second generation of controllers is beg to emerge. About a dozen thyristor-based systems have been identified as likely to improve the performance of an AC system. Six are being pursued for development as part of EPRI's proposed 10-year collaborative R&D plan for the technology.

Obviously, most vital to power and stability control is the ability to control impedance or phase angle. Since the series impedance of a typical transmission line is mostly inductive, with only 5%-10% resistive, it is convenient to control a power system's steady-state impedance by adding both a thyristor-controlled series capacitor and a thyristor-controlled series reactor (inductor). Since the capacitor is a negative impedance, the introduction of a variable series capacitor means a variable negative impedance in series with the line's natural positive impedance.

5. Thyristor-Controlled Series Capacitor

Thus the thyristor-controlled series capacitor can vary the impedance continuously to levels below and up to the line's natural impedance. On the other hand, adding a thyristor-controlled series reactor means adding a variable positive impedance to a value above the line's natural positive impedance. Once installed, it will either respond rapidly to control signals to increase or decrease the capacitance or inductance, thereby damping those dominant oscillation frequencies that would otherwise breed instabilities or unacceptable dynamic conditions during and after a disturbance.

The first of the new controllers to be demonstrated on a utility transmission system is the thyristor-controlled series capacitor (TCSC). In 1991, American Electric Power Co. of Columbus, Ohio, began testing a prototype switch on one phase of the series capacitor bank at its 345 kV Kanawha River Substation in West Virginia. The switch was supplied by Asea Brown Boveri, Vasteras, Sweden. Although the switch is not strictly speaking a thyristor controlled series capacitor and the test installation is only single phase, the rapid switching of capacitor segments in and out of the circuit do test the key hardware. In some applications, the thyristor-based switching is justifiable for its reliability and speed.

In October 1992, the Western Area Power Administration (WAPA), Golden, Colo., dedicated the first three-phase thyristor-controlled series capacitor installation. It was built by Munich-based Siemens AG and installed at the WAPA Kayenta Substation in Arizona in a 300 km, 230 kV, 300 MW transmission line. The installation includes three segments: one 15 Ω, one 40 Ω, and one 55 Ω series capacitor bank. Only the 15 Ω (45 MV·A) bank is thyristor controlled. This installation is a pioneering step because the 15 Ω bank can be controlled smoothly and rapidly from 15 Ω to 60 Ω through the controlled firing angle of the thyristor valves as in the NGH SSR damper, but over a wider range, from 145° to 180°. The installation allows the transmission line capacity to be increased from 300 MW to 400 MW.

The construction of a large three-phase thyristor-controlled series capacitor installation on a 500 kV line is in progress at the Slatt Substation of Bonneville Power Administration (BPA) in Oregon (this

BPA line is connected to Portland General Electric Co.'s Boordman power plant in Oregon). The installation features a full range of controls and operating requirements.

Developed and designed by GE in Schenectady, New York, with EPRI funding, it is due for completion by mid-year. Each of the three-phases includes a platform that consists of six identical series-connected capacitor modules, 1.33 Ω each, with a parallel-connected thyristor switch in series with 0.2 Ω inductor (Fig. 6.3). Each module can be controlled smoothly by advancing the firing angle by a few degrees from 180° of each half wave, as in the NGH-SSR damper. Also, when the thyristor in a capacitor module is fully conducting, the equivalent of a parallel connected −1.33 Ω capacitor and a +0.2 Ω inductor gives an impedance of +0.24 Ω. Thus with a combination of a pull bypass and partial conduction of all six modules, the impedance can be controlled approximately from +14 Ω to −16 Ω (the larger value is obtainable with certain firing-angle combinations).

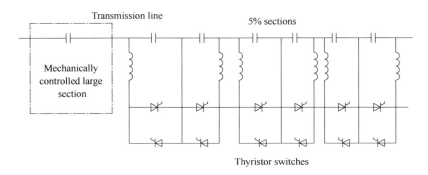

Fig. 6.3 Thyristor-controlled series capacitor

The installation has a rated current of 2900 A, with a short-term overload rating of 4350 A for 30 minutes and 5900 A for 10 seconds, a fault current withstand rating of 20000 A (rms).

Each thyristor switch consists of five thyristors in series, triggered by light signals from the control room. Deionized water passes through insulating hoses to cool the thyristors, maintaining an ambient temperature range of −40℃ to +40℃.

The thyristor-controlled series capacitor includes remote impedance control, power control, and current control through a SCADA system. Functional features of the NGH-SSR are included to mitigate the sub-synchronous resonance, not to mention power swing damping, transient stability control, assorted local protection features and overload management. The modularity and high current rating of this installation will confirm this technology's usefulness for a whole range of requirements.

The Bonneville Power Administration is also studing the application of the thyristor-controlled series capacitor in specific AC interties, namely, for transmission lines crossing the Cascade Mountains to the areas around Puget Sound and Seattle, Wash and around Portland, Ore, and also to the Western Montana transmission system. All have potential for the enhancement of usable thermal capacity.

6. Static Condenser

As mentioned earlier, the static var compensator that uses thyristor switches is already a firmly established piece of equipment for voltage control. In the years to come, however, it will be outperformed by another concept called the static condenser, or STATCON (Fig. 6.4a). In essence, The STATCON is a three-phase inverter that is driven from the voltage across a DC storage capacitor and whose three output voltages are in phase with the AC system voltages. When the output voltages are higher (or lower) than the AC system voltages, the current flow is caused to lead (or lag), and the difference in the voltage amplitudes determines how much current flows. In this manner, reactive power and its polarity can be controlled by controlling the voltage.

Why the superior performance? The most reactive power that can be delivered from the STATCON equals the voltage times the current, whereas in the case of its predecessor, the SVC, it is the square of the voltage divided by the impedance. Consequently, if the voltage is depressed, the STATCON can still deliver high levels of reactive power by using its overcurrent capability. With the SVC, on the other hand, the reactive power capability falls off steeply as a function of the square of the voltage-just when it is needed most. Thus, depending on the application, the rating of the STATCON required will be much smaller than the rating of a comparable SVC. In addition, a Statcon equipped with a large DC capacitor or large storage device, such as a battery bank or superconducting storage reactor, can continue to deliver some energy for a short time, just as a synchronous condenser does because of the energy stored in its rotating mass.

However, the STATCON does require gate turn-off thyristors, which at present cost more than ordinary thyristors yet have higher losses and lower voltage and current ratings. Thus, ordinary thyristor technology has achieved a rating of 8 kV, 4000 A (32 MW peak switching per device), while the gate turn-off technology has achieved 4.5 kV, 3000 A (13.5 MW peak switching per device). Furthermore, the most promising thyristor concept with gate turn-off ability is the MOS-controlled thyristor (MCT), pioneered through EPRI funding. It is already commercialized at low rating, but has yet to be developed on a scale suitable for utility application.

An experimental 1 Mvar STATCON developed by EPRI with Westinghouse was demonstrated at Orange & Rockland Utilities Inc. in New York State trough the sponsorship of Empire State Electric Energy Research Corporation, headquartered in New York City. Now, with the sponsorship of EPRI and the Tennessee Valley Authority (TVA as sponsors of Westinghouse), a 100 Mvar STATCON is due to be demonstrated in 1995 at the Knoxville-based TVA's Sullivan Substation. This unit, apart from controlling the voltage, will also be used to damp power oscillations. Without the STATCON, TVA might have had to construct another transmission line. Another way to control the power flow on the transmission line is the phase angle regulator, as also noted earlier.

7. Phase Angle Rebulator

One of a number of different concepts for it is shown in Fig. 6.4b. The phase shift is accomplished by adding or subtracting a variable voltage component that is perpendicular to the phase voltage of the line. This perpendicular voltage component is obtained from a transformer connected between the other two phases. In the scheme shown, the three secondary windings have voltages proportional to 1 : 3 : 9. Thyristor switches, one per winding, allow each winding to be included or excluded in the positive or negative direction. The choice of 1%, 3%, and 9% along

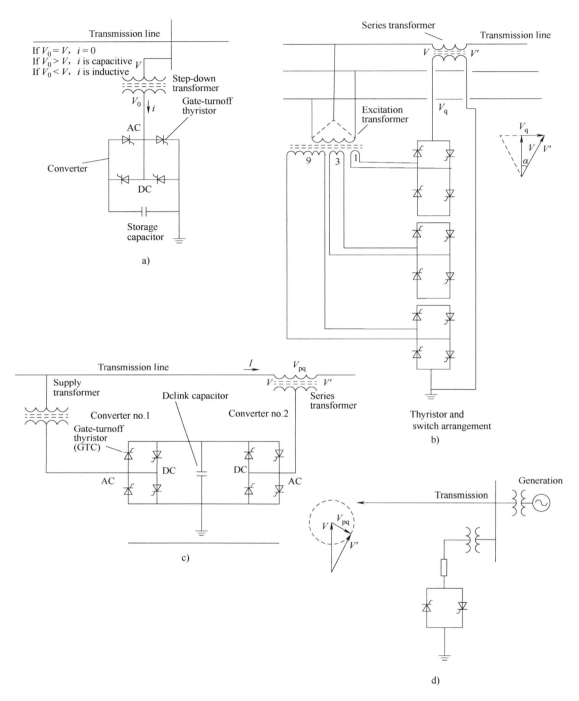

Fig. 6.4 Concepts of phase angle regulator

with the plus or minus polarity for each winding yields a switchable voltage range of -13-$+13$ V, thus giving a variable high speed control of the perpendicular voltage component. The voltage corresponding to each unit step will of course determine the total phase shift that results.

In Fig. 6.4, functionally equivalent to a synchronous condenser, a STATCON is an inverter typically based on gate turn-off thyristors (Fig. 6.4a). The inverter supplies capacitive or inductive reactive power depending on whether the storage capacitor's voltage V_0 is larger or smaller than the line's voltage V. A conceptual thyristor-controlled phase angle regulator employs an excitation transformer with three secondary windings with turns ratios $1:3:9$; the thyristor switch produces a voltage V_q that is added in series to the line voltage V to produce a phase shift α (Fig. 6.4b)

A unified power controller injects an AC voltage V_{pq} in series with the line voltage V, there-by allowing the control of the phase angle between the resultant line voltage V' and the current i (Fig. 6.4c). A dynamic brake (a thyristor-controlled resistor in parallel with the transmission line) can be used effectively to damp power-swing oscillations in the transmission system (Fig. 6.4d).

In another concept of interest, known as the unified power flow controller (Fig. 6.4c), an AC voltage vector generated by a thyristor-based inverter is injected in series with the phase voltage. The driving DC voltage for the inverter is obtained by rectifying the AC to DC from the same transmission line. In such an arrangement, the injected voltage may have any phase angle relationship to the phase voltage. It is therefore possible to obtain a net phase and amplitude voltage change that confers control of both active and reactive power.

EPRI is studying these concepts with WAPA and plans to hold demonstrations of them in the coming years. Generally, the impedance control would cost less and be more effective than the phase angle control, except in the condition that the phase angle is very small or very large or varies widely.

8. Other Controllers

The dynamic brake (Fig. 6.4d) is a shunt connected resistive load, controlled by thyristor switches. Such a load can be, generally, the impedance control selectively applied in each pass, half cycle by half cycle, to damp any specific power flow oscillation, so that generating units run less risk of losing synchronism. As a result, more power can be transferred over systems subject to stability constraints.

The dynamic voltage limiter is a shunt-connected, high-power zinc oxide gap-less arrestor controlled by thyristor switches. It can be used to limit over voltages for hundreds of milliseconds if the transmission line capacity is affected by high dynamic over voltages.

The thyristor-controlled series reactor resembles the thyristor-controlled series capacitor in that it has a thyristor switch connected across the reactors for swift control of the line's effective inductive impedance.

In all probability, other controllers will be invented. More likely still, numerous improvements will be made in the effectiveness and cost of the basic controllers. Tab. 6.1 opposite lists the key controllers along with their attributes.

Tab. 6.1 Functions of flexible AC transmission system controllers

Types	Attributes
NGH-SSR damper	Damping of oscillations, series impedance control, transient stability
Static var compensator (SVC)	Voltage control, var compensation, damping of oscillations

（续）

Types	Attributes
Thyristor-controlled series capacitor	Power control, voltage control, series impedance control, damping of oscillations, transient stability
Static condenser (STATCON)	Voltage control, var compensation, damping of oscillations, transient stability
Thyristor-controlled phase angle regulator	Power control, phase angle control, damping of oscillations, transient stability
Unified power flow controller	Power control, voltage control, var compensation, damping of oscillations, transient stability
Thyristor-controlled dynamic brake	Damping of oscillations, transient stability

NGH-SSR = Narain G. Hingorani-subsynchronous resonance; var = vlot ampere reactive (power).

9. Maximizing Capacity

Long delays and nearly insurmountable problems often hamper or stall the job of securing new rights-of-way or building or rebuilding lines on existing rights-of-way. Yet additional transmission is often required to reduce the need for new generation, shrink generation reserve margins and improve the fuel economy, all of which are benefit to the environment. For a long time, utilities have faced the so-called loop flow problem, and now the heavy pressure for access to transmission lines will create many new demands on both line capacity and the ability to control power flow. Hence the timeliness of flexible AC transmission system technology, can ensure that power flows through the prescribed routes, that the capacity of existing corridors is maximized, and that the secure loading capability is increased under various scenarios of upgrading or upgrading the lines' thermal current capacity. EPRI is also investigating transmission line upgrading configurations, that will increase transmission capacity as well as decrease the magnitude of magnetic fields.

Flexible AC transmission system technology opens up new utility planning scenarios putting on the agenda a whole array of options consistent with a new age of competition and environmental considerations. Although this makes the planning process more complex and challenging, planners are responding with excitement and enthusiasm.

Therefore, it is important that planners have the appropriate tools. EPRI funded planning studies by GE and Power Technologies Inc., Schenectady, New York, both of which have enhanced their computer codes to incorporate some flexible AC transmission system controllers. EPRI's own comprehensive code package, PSAPAC, is continually being updated with more detailed simulations of all the new controllers. These planning tools must continue to be expanded and improved, for only then will the simulation of various controllers be sophisticated enough for planners to consider flexible transmission options routinely and not as an afterthought. There are now eight EPRI member utilities engaged in planning studies investigating flexible AC transmission systems as an option in depth.

Interestingly enough, for any given case study related to increasing usable transmission capacity, two planners will most likely come up with different scenarios. That is only to be expected when planning is made more complicated by greatly expanded options.

Much work still needs to be done to firmly establish flexible transmission technology, the most important being the commercial demonstration of various controllers. Their applicability will expand as each new type becomes available. In essence, the controllers are ingenious combinations of advanced but commercially available components, with the thyristor having the most influence on cost.

Another aspect of application is the system and engineering cost. Flexible ac transmission applications currently involve considerable customization. However, as the market grows, consolidation of modular designs and marked cost reductions can be expected. A third issue is incentives for new capital investments in the technology, to accommodate new transmission access needs.

Systems engineering as well as hardware aspects are being studied by working groups in the IEEE and the International Conference on Large High Voltage Electric Systems (Cigré) at present.

Within the next 20 years, with any luck, electricity should be flowing through controllers between many generating plants and customers, optimizing the use of utility transmission systems.

6.2 Static Var Compensator (SVC)

For many years, static var compensators have been an essential component in the operation of power transmission systems. They are part of a family of devices known as FACTS devices. The advent of large capacity force-commutated semiconductor switches allows many developments in power electronic converters to be applied to the implementation of high power compensators. This section describes the principles of controlled reactive power compensation, particularly in the context of power systems. It focuses on active static power converter-based compensators and discusses issues related to the power circuit topology and control techniques, including the impact of pulse width modulation (PWM) techniques. Compensators based on current, voltage source converters and on AC controllers, both in the shunt and series configurations, are covered. Methods to enhance power capacity using multi-level and multi-pulse arrangements are discussed.

6.2.1 Introduction

Reactive power (var) compensation has long been recognized as an essential function in the operation of power systems. At the distribution level, it is used to improve the power factor and support the voltage of large industrial loads, such as line commutated thyristor drives and electric arc furnaces. Reactive power compensation also plays a crucial role at the transmission level in supporting the line voltage and stabilizing the system. Rotating synchronous condensers and mechanically-switched capacitor and inductor banks have been replaced in the 1970s by thyristor-based technologies: in typical installations, a TCR provides variable lagging vars, and fixed or TSC provides the leading vars. The combination of both devices in parallel allows continuous control of vars over a wide range of values, from leading to lagging vars. A large number of units have been successfully installed and operated for many years. At the same time, the potential of var compensators based on static power converters have also been recognized and a number of configurations proposed and investigated. However, thyristor technology only allows the implementation of lagging var generators, unless complex force-commutation circuits are used. This drawback has been eliminated with the introduction of GTO thyristors. This has allowed the development of a number of configurations based on the use of synchronous voltage sources. Prototype GTO-based var compensation units, known as STATCOMs have been installed and tested by utilities. The STATCOM and other static var compensators have recently been grouped, together with other types of transmission system control devices, under the heading of FACTS devices. Reactive power compensators are typically connected in shunt across transmission and distribution systems. An alternative connection, the series connection, has recently received much attention from utilities. Technological solutions have been developed to solve problems associated with insulating the equipment from ground and the full potential of series connections can now be exploited. The latest development in var compensation technology has been the combination of series and shunt static compensators into one unit, known in the area of power systems under the name of unified power flow controller, or UPFC.

Static power converters have been successfully applied to a large number of power conversion problems at low and medium power levels. However adapting these solutions to high power transmission and distribution levels raises special issues. Although the capacity of power semiconductor switching devices has gradually increased, large ratings still require combining devices in series and parallel. In addition to the large power handling capacity, static compensators must have very high efficiency, since losses have a negative impact on both the initial and operating costs of the power system. Switching losses are therefore a primary concern and switching frequencies must therefore be kept low. This may result in large harmonic waveform distortion, unless special power circuit configurations are used.

This section reviews the various methods available for generating reactive power (var) by means of force-commutated static power converters, taking the above constraints into account. It discusses topologies suitable for use with devices such as GTOs and the more recently available high power IGBTs and addresses switch gating issues, including the use of PWM techniques. Methods for designing high power converters suitable for transmission level compensation are presented, particularly multilevel and multi-module topologies.

6.2.2 Principles of Var Compensation

Var compensation can be viewed as the injection of reactive power, leading or lagging into the AC system. In its simplest form, reactive power injection is achieved by inserting fixed capacitors or inductors in either series or shunt into the AC system. Assuming a compensation reactance X_c is inserted in a transmission system, the generated var Q_c is derived as follows.

1) For the shunt connection, in Fig. 6.5a, a reactive current I_{cq} is generated, allowing in particular line voltage support at the point of connection, V_c:

$$I_{cq} = V_c / X_c$$
$$Q_c = V_c^2 / X_c$$

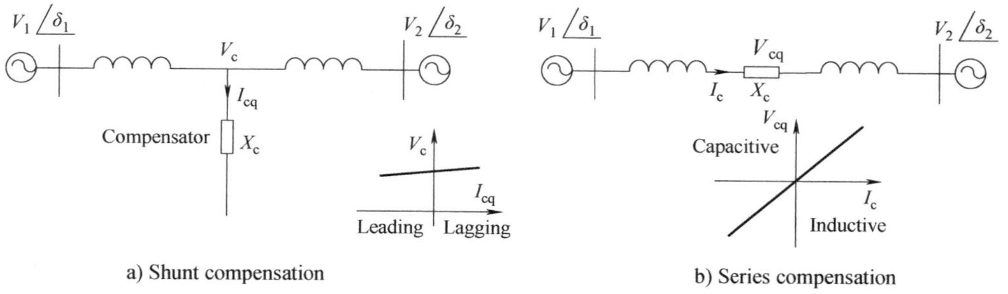

Fig. 6.5 Principle of var compensation in transmission systems

2) For the series connection, in Fig. 6.5b, the reactive impedance X_c partially compensates the line reactance, and a reactive voltage V_{cq} is inserted in series, the current I_c is the line current:

$$V_{cq} = I_c/X_c$$
$$Q_c = I_c^2/X_c$$

In addition to the lack of controllability of the reactive power injection, fixed capacitive compensation can lead to AC system instability, such as in the phenomenon known as subsynchronous resonance, or SSR associated with series compensation. In order to control the amount of reactive power injected, the reactive impedance must be varied. Equivalently, a variable current or variable voltage is injected into the system, emulating a variable reactance, as Fig. 6.5b shows. The apparent reactive impedance of a fixed element can be varied using AC switches, or AC controllers. On the other hand, the current or voltage required to emulate a variable reactance can be injected into the AC system by means of synthetic sources, which can be realized using static power converters. In addition to providing reactive power, these active compensators can also supply real power, either transiently or for a number of periods of the AC supply. This real power can be used to dampen power system oscillations or temporarily support the power system voltage under fault conditions. Furthermore, since the compensator is fully controllable, resonant frequencies associated with the use of capacitors are eliminated and the potential for instability suppressed.

6.2.3 AC Controller-Based Structures

1. Conventional Thyristor-Controlled Reactor

The basic scheme, in Fig. 6.6a, consists of an AC controller which varies the apparent inductance of the inductor as reflected on the AC line. The TCR provides continuously controllable lagging vars and is biased using fixed, or more often, TSC. The injected vars can therefore be continuously adjusted from leading to lagging, as shown in Fig. 6.6b. However, the var injection or voltage regulation capability of the compensator is limited by the value of the reactance and is therefore line voltage dependent, as shown in Fig. 6.6b. Advantages of the system include ruggedness, high efficiency, good dynamic performance and competitive cost. The disadvantages include injection into the line of large low frequency harmonic currents, dominant being the 5th and 7th components (300 Hz and 420 Hz for a 60 Hz system) for the basic AC controller. Harmonics can be moved to higher frequencies by paralleling units and using special transformer configurations. Harmonic currents can be reduced by means of tuned LC filters. However these are costly and can cause voltage oscillations resulting from the added system resonant frequencies.

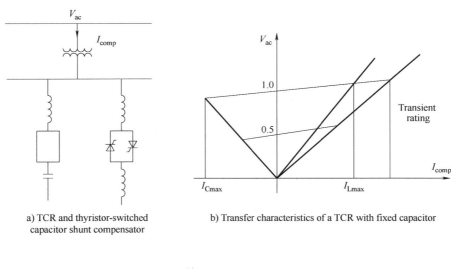

a) TCR and thyristor-switched capacitor shunt compensator

b) Transfer characteristics of a TCR with fixed capacitor

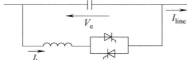

c) Thyristor-controlled series capacitor compensator (TCSC)

Fig. 6.6 Thyristor-controlled variable reactance compensators

2. Thyristor Series Controlled Capacitor

The dual of the TCR is the thyristor-controlled series capacitor (TCSC), shown in Fig. 6.6c. It uses a thyristor-based AC controller to adjust the apparent reactance seen from the line continuously. Such units have been installed in transmission systems successfully. However, they have the same limitations as TCRs, including harmonic injection. Performance and harmonic reduction may be enhanced by using force-commutated switches. Although AC controller-based compensators are rugged and simple to control, operating regions are determined by the reactance used and the line conditions (Fig. 6.6b). Converter-based structures are more versatile and flexible, and are therefore receiving more attention.

3. Force-Commutated AC Controller Structures

An alternative to the thyristor-based AC controllers is the force-commutated AC controller, as shown in Fig. 6.7a. The use of force-commutated switching devices removes the requirement for operating the converter in synchronism with the AC supply and allows gating the switches more than once per cycle. Arbitrary gating particularly PWM patterns can be implemented. Assuming the inductor current is sinusoidal, a pattern with constant duty cycle yields AC line side currents that only contain harmonics around the switching frequency and its multiples. This pattern is simple to

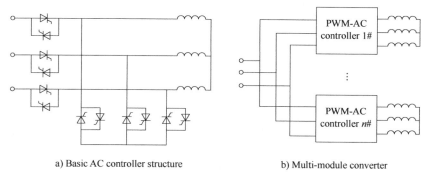

a) Basic AC controller structure   b) Multi-module converter

Fig. 6.7 Variable reactance scheme based on fore-commutated AC controllers (bias capacitors are not shown)

implement and allows control of the equivalent inductance; therefore, the amount of vars absorbed can be varied from 0 to a maximum value. Since losses associated with switching large currents at high voltages increase significantly with switching frequency, this frequency must be kept low, typically a few hundred hertz for GTOs. In order to reduce the distortion of injected currents, while keeping the switching frequency low, elementary modules are connected in parallel (Fig. 6.7b), and gated so that harmonics are minimized. This also allows the realization of var compensators of large ratings.

6.2.4 Active Compensator Topologies

The availability of controllable voltage and current sources, realized using static power converters, allows the implementation of static compensators which can emulate the characteristics of a variable reactance. Advantages over the variable reactance structures include the following.

1) The performance characteristics, especially reactive power generation capability are improved, and are independent of the line parameters and line steady state operating voltages and currents.

2) Energy storage devices, such as battery banks and superconducting magnet energy storage (SMES) systems can be incorporated into the units. This allows the absorption and generation of real power, which can be used for power system damping and voltage support under fault conditions. In the shunt connection, the power converter system behaves in a manner similar to a synchronous machine operated as a synchronous condenser. If the converter is considered a voltage source behind an inductance, varying the output voltage is equivalent to varying the internal voltage of the synchronous generator.

6.3 Uniform Power Flow Controller (UPFC)

6.3.1 Introduction

The control of an AC power system in real time is involved because power flow is a function of the transmission line impedance, the magnitude of the sending and receiving end voltages, and the phase angle between these voltages. Years ago, electric power systems were relatively simple and were designed to be self-sufficient; power exportation and importation were rare. Furthermore, it was generally understood that AC transmission systems could not be controlled to handle dynamic system conditions fast enough. Transmission systems were designed with fixed or mechanically-switched series and shunt reactive compensations, together with voltage regulating and phase-shifting transformer tap-changers, to optimize line impedance, minimize voltage variation and control power flow under steady-state or slowly changing load conditions. The dynamic system problems were usually handled by over design; transmission systems were designed with generous stability margins to recover from anticipated operating contingencies caused by faults, line and generator outages, and equipment failures. All these resulted in the (often considerable) underutilization of transmission systems.

In recent years, energy, environment, right-of-way, and cost problems have delayed the construction of both generation facilities and new transmission lines, while the demand for electric power has continued to grow. This situation has necessitated a review of the traditional power system concepts and practices to achieve greater operating flexibility and better utilization of existing power systems.

During the last two decades, major, if not revolutionary, advances have been made in high-power semiconductor device and control technologies. These technologies have played a great role in the broad application of HVDC transmission and power system inertia schemes, and they have already made a significant impact on AC transmission via the increasing use of thyristor-controlled static var compensators (SVCs).

Static var compensators control only one of the three important parameters (voltage, impedance, phase angle) determining the power flow in AC power systems: the amplitude of the voltage at selected terminals of the transmission line. Theoretical considerations and recent system studies indicate that high utilization of a complex, interconnected AC power system, meeting the desired objectives for availability and operating flexibility, may also require the real-time control of the line impedance and the phase angle. Hingorani proposed the concept of FACTS, which includes the use of high-power electronics, advanced control centres, and communication links, to increase the usable power transmission capacity to its thermal limit. Within the framework of FACTS and other efforts with similar objectives, the development of thyristor-controlled series compensators for line impedance control, thyristor-controlled tap-changing transformers for phase angle control, and other thyristor-controlled devices for dynamic 'brakes' and overvoltage suppressors has already been started or is expected to start in the near future.

Although present static var compensators and other thyristor-controlled equipments developed for power flow control (i.e., series compensators and phase shifters) can have the necessary speed for real-time control, they are rather large, custom-designed and fabricated systems of substantial cost, requiring considerable size facility with significant labor installation. For these reasons, it is unlikely that to provide a long-term, volume-production based economic solution for flexible AC transmission systems.

It has long been realized that an all solid-state or advanced, static var compensator, which is the true equivalent of an ideal synchronous condenser, is feasible technically. And with the use of gate turn-off (GTO) thyristors, is economically viable. The extension of this approach to controllable series compensation and phase shifting has been proposed recently. This uniform approach of power-transmission control promises simplified system design, reduction in equipment size and installation labor, improvements in performance, and significant reduction in capital cost, fuelled by advances in power semiconductor technology.

The objective of this section is to outline the technical and economical factors which characterize the uniform, all solid-state power-flow controller approach for real-time controlled flexible AC transmission systems.

6.3.2 Basic Structure of the Unified Power Flow Controller

The unified power flow controller (UPFC) was proposed for real-time control and dynamic compensation of AC transmission systems, providing the necessary functional flexibility required to solve many of the problems facing the utility industry.

The UPFC consists of two switching converters, which are voltage-sourced inverters using GTO thyristor valves in the implementations considered, as illustrated in Fig. 6.8. These inverters, labeled as "Inverter 1" and "Inverter 2" in the figure, are operated from a common DC link provided by a DC storage capacitor. This arrangement functions as an ideal AC to AC power converter in which the real power can freely flow in either direction between the AC terminals of the two inverters and each inverter can independently generate (or absorb) reactive power at its own AC output terminal.

Inverter 2 provides the main function of the UPFC by injecting an AC voltage v_{pq} with controllable magnitude V_{pq} ($0 \leq V_{pq} \leq V_{pqmax}$) and phase angle σ ($0 \leq \sigma \leq 360°$), at the power frequency, in series with line via an insertion transformer. This injected voltage can be considered as a synchronous AC voltage source essentially. The transmission line current flows through this voltage source resulting in real and reactive power exchange between it and the AC system. The real power exchanged at the AC terminal (i.e., at the terminal of the insertion transformer) is converted by the inverter into DC power which appears at the DC link as positive or negative real power demand. The reactive power exchanged at the AC terminal is generated by the inverter internally.

The basic function of Inverter 1 is to supply or absorb the real power demanded by Inverter 2 at the common DC link. This DC link power is converted back to AC and coupled to the transmission line via a shunt-connected transformer. Inverter 1 can also generate or absorb controllable reactive power, if it is desired, thereby it can provide independent shunt reactive compensation for the line. It is important to note that whereas there is a closed "direct" path for the real power negotiated by the action of series voltage injection through Inverters 1 and 2 back to the line, the corresponding reactive power exchanged is supplied or absorbed by Inverter 2 locally and therefore it does not flow through the line. Thus, Inverter 1 can be operated at a unity power factor or be controlled to have a reactive power exchange with the line independently of the reactive power exchanged by Inverter

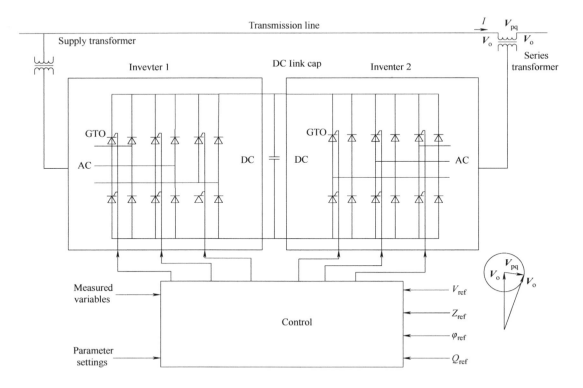

Fig. 6.8 Basic circuit arrangement of the united power flow controller

2. This means that there is no continuous reactive power flow through the UPFC.

Viewing the operation of the UPFC from the standpoint of conventional power transmission based on reactive shunt compensation, series compensation, and phase shifting, the UPFC can fulfill all these functions and thereby meet multiple control objectives by adding the injected voltage V_{pq} with appropriate amplitude and phase angle, to the terminal voltage V_o. Using phasor representation, the basic UPFC power flow control functions are illustrated in Fig. 6.9.

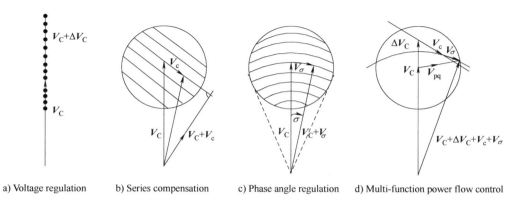

a) Voltage regulation   b) Series compensation   c) Phase angle regulation   d) Multi-function power flow control

Fig. 6.9 Basic UPFC control functions

6.4 Bridge-Type Fault Current Controller (FCC)

6.4.1 Introduction

The bridge-type fault current controller (FCC), previously called fault current limiter (FCL), was introduced in 1983. For a single-phase circuit the controller consists of a full-wave bridge, an inductance, and an optional bias power supply. Operation of the controller is based on providing a bias current through four thyristors. This bias current allows an inductor to be inserted into the load circuit when the load current exceeds the bias current. For load currents smaller than the bias current, the forward biased thyristors present a zero-impedance path to the load current. Should the load current exceed the bias current, as under short circuit conditions, a pair of thyristors is turned off during each half cycle and the inductance is automatically switched into the circuit, providing current limiting. Because phase control can be applied to the thyristors, the AC current can be regulated in magnitude.

The name FCL was changed to FCC after it was recognized that the controller has more attractive features besides limiting fault currents. The two main additional features are the capability to act as a variable inductance and as a less-than-a-cycle solid-state circuit breaker. The initial version of the FCC required a bias power supply for proper operation of the controller. It was found later that the bias power supply is unnecessary, the bias current could be provided by the source by proper thyristor control. The complexity of the controller operating without the bias power supply was reduced considerably without compromising the main features of the unit.

As is the case with other FACTS controllers, which consist of high power solid-state switching devices and passive components, the FCC uses the same devices. The FCC also controls parameters of the power system and in that sense the FCC is a new FACTS controller.

This sector explains the operation of the controller with and without a bias power supply. Unique features of the FCC will be pointed out. Test results of a 13.7 kV unit, operating in a single-phase mode, will be presented. The test results are in close agreement with circuit simulation results. Potential applications of the controller in utility and industrial power systems will be outlined.

6.4.2 Bridge-Type Fault Current Controller with a Bias Power Supply

The original version of the FCC is shown in Fig. 6.10 for a single-phase circuit, having a voltage source V_s and a source impedance Z_s. The FCC, consisting of four thyristors (1, 2, 3, 4) in a bridge-circuit arrangement, a coil L_b with a resistance R_b, and a DC (bias) power supply V_b, is connected in series between the source and the load. The bias power supply establishes a DC current through the four thyristors and the coil. Under no load conditions, each thyristor conducts half of the DC coil current. Under load conditions, with the load current smaller than the bias current, the AC current will be superimposed on the DC bias current in the thyristors. For example, during the half cycle with positive current flow, the AC current will add to the DC bias current in thyristors 1, 2, and subtract in thyristors 3, 4. The current through each thyristor remains positive as long as the amplitude of the AC current is smaller than the DC bias current. The AC current is only limited by the load impedance, if we assume lossless thyristors. If the load impedance is lowered, as is the case during a short circuit, and the amplitude of the AC current tries to exceed the DC bias current, one of the two thryristor sets (1, 2 or 3, 4) will go into the blocking state when the thyristor current is reduced to zero. Positive current flow thyristors 1, 2 and for negative current flow thyristors 3, 4 will conduct. For load currents higher than the bias current, the inductance of the FCC is switched into the circuit and limits the AC current. Once a short circuit is initiated, each thyristor conducts current every half cycle. If a controlled bias voltage source is used, the bias current level can be adjusted to set the load current level at which the FCC inductance is switched into the circuit.

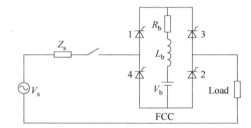

Fig. 6.10 Diagram of a single-phase, bridge-type fault current controller with a bias power supply, installed between source and load

There are several features of the FCC that are worth mentioning:

1) The first half cycle of the short-circuit current is already limited in amplitude, because the inductance of the FCC will be switched into the AC circuit at the moment the fault current is higher than the bias current. The maximum value of current limiting is determined by the size of the FCC inductance.

2) The FCC inductance is switched into the AC circuit automatically by two sets of blocking thyristors. No active sensing mechanism is necessary, which should result in a reliable controller.

3) Phase-delay angle control of the thyristors allows to adjust of the short-circuit current after the first current half-cycle. Changing the phase-delay angle α from 0° to 90° results in a gradual decrease of the short-circuit current. The AC current has a sinusoidal shape for phase angles up to 90°. The short-circuit current will decrease further for angles greater 90°. At $\alpha = 90°$ the effect of the FCC is the same as if it were replaced by a series connected inductor L_b. For angles $\alpha > 90°$ the FCC produces AC currents that are discontinuous and adjustable in amplitude. This feature of varying the amplitude of the line current justifies the name "current controller" instead of simply "current limiter."

4) For load currents higher than the bias current, two sets of thyristors always alternate in conducting the current. If the gate pulses to the thyristors cease, the FCC acts as a circuit breaker and interrupts the current. In the worst condition when the short circuit happens after a particular set

of thyristors fire, gate blocking results in a current interruption time of about one cycle for an inductive circuit. This current interruption time can be improved to about one-half cycle by a modified firing scheme.

In principle, three of these controllers are necessary for a three-phase system. In a delta connected three-phase system, two single-phase controllers suffice.

Because the FCC coil conducts the full bias current during operation at all times, the possibility of using a superconducting coil to reduce the steady-state losses of the unit is been pursued.

VOCABULARY

1. thermal adj. 热的，热量的
2. thrust n. 推力；刺
3. compensator n. 补偿器；自耦变压器
4. flywheel n. 飞轮，惯性轮；调速轮
5. margin n. 界限；差数；幅度
6. oscillation n. 振荡；振动；摆动
7. bulk n. （巨大的）体积，容积，容量，（大）量
8. alleviate vt. 减轻，缓和
9. meshed adj. 网状的；有网孔的
10. shaft n. 杆；轴
11. turbine n. 涡轮；涡轮机
12. detune vt. 使去谐；解谐；失谐
13. collaborative adj. 合作的，协作的
14. prototype n. 原型；标准，模范
15. deionize vt. 除去离子
16. ambient adj. 周围的；环绕的；外界的
17. mitigate vt. 使缓和，使减轻
18. modularity n. 模块性
19. intertie n. （美）连锁电力网
20. steeply adv. 陡峭地；险峻地
21. perpendicular adj. 垂直的，正交的；直立的
22. rectify vt. 整流
23. demonstration n. 示范，展示
24. hamper vt. 妨碍；束缚
25. stall vt. 拖延；使停转
26. shrink vt. 使缩小，使收缩
27. crucial adj. 重要的；决定性的；定局的
28. distortion n. 变形；扭曲；失真
29. topology n. 拓扑结构，拓扑
30. temporarily adv. 临时地，临时
31. ruggedness n. 强度，坚固性
32. sinusoidal adj. 正弦曲线的
33. fabricate vt. 制造；装配
34. superimposed adj. 叠加的；上叠的；重叠的
35. delta n. 三角形物

NOTES

1. Existing installations that use line-commutated thyristor based technologies are either static var compensators (SVC) or thyristor controlled series compensators (TCSC).
现有的静态无功补偿（SVC）或者可控串联补偿（TCSC）装置使用的都是电网换相的晶闸管技术。

2. The development of flexible transmission system is akin to high-voltage DC and related thyristor, designed to overcome the limitations of the present mechanically controlled AC power transmission systems.
柔性交流输电系统的发展类似于高压直流输电及其相关的晶闸管技术，旨在克服目前机

械控制的交流输电系统的局限性。

3. Granted, mechanical controllers are on the whole less expensive, but they increasingly need to be supplemented by rapidly responding power electronics controllers.

诚然，机械投切控制总体相对便宜，但它们越来越需要快速响应的电力电子控制器的辅助作用。

4. A well-known formula states that the power flow between two points along a transmission line is equal to the product of the voltages at these points, times the sine of the difference between their phase angles, all divided by the transmission line's reactance between the two points.

一个众所周知的公式表明，输电线路两点之间的功率等于两点电压的积，乘以两点相位差的正弦值，再除以两点之间传输线的电抗。

5. If, however, a capacitor whose reactance is $-5\ \Omega$ at the synchronous frequency is inserted in one line (Fig. 6.1b), it reduces the line's impedance reduces from $10\ \Omega$ to $5\ \Omega$, so that power flowing through the lines will be 250 MW, 1250 MW, and 1750 MW, respectively.

但是，如果在一条线路上插入一个同步频率时电抗为 $-5\ \Omega$ 的电容（Fig. 6.1b），则线路阻抗从 $10\ \Omega$ 减少为 $5\ \Omega$，所以流过线路的功率相应分别为 250 MW、1250 MW 和 1750 MW。

6. The NGH-SSR damper consists of a thyristor AC switch (back-to-back thyristors) connected in series with a small inductor and resistor across the series capacitor.

NGH-SSR 阻尼器由一个晶闸管交流开关（双向晶闸管）串联一个小电感和电阻后再与一个串联电容器并联构成。

7. Since the series impedance of a typical transmission line is mostly inductive, with only 5%-10% resistive, it is convenient to control a power system's steady-state impedance by adding both a thyristor-controlled series capacitor and a thyristor-controlled series reactor (inductor).

由于一个典型的输电线路的串联阻抗主要是感性的，只有5%-10%是阻性的，因此可以很方便地通过增加晶闸管控制的串联电容器和晶闸管控制的串联电抗器来控制电力系统的稳态阻抗。

8. Functional features of the NGH-SSR are included to mitigate the sub-synchronous resonance, not to mention power swing damping, transient stability control, assorted local protection features and overload management.

NGH-SSR 的功能特性包括降低次同步谐振、阻尼功率振荡、暂态稳定控制、就地保护功能和负荷过载管理。

9. When the output voltages are higher (or lower) than the AC system voltages, the current flow is caused to lead (or lag), and the difference in the voltage amplitudes determines how much current flows.

当输出电压高于（或低于）交流系统电压，会造成电流超前（或滞后），电压幅值差决定电流的大小。

10. In another concept of interest, known as the unified power flow controller, an AC voltage vector generated by a thyristor-based inverter is injected in series with the phase voltage.

另一个有趣的概念如统一潮流控制器，（通过串联变压器）将晶闸管逆变器所产生的交流电压矢量串联加到相电压上。

11. For a long time, utilities have faced the so-called loop flow problem, and now the heavy pressure for access to transmission lines will create many new demands on both line capacity and the ability to control power flow.

电力公司在过去很长一段时间里一直面对所谓的闭环潮流问题，如今传输线通道的巨大压力对线路容量和潮流控制能力提出了新的要求。

12. On the other hand, the current or voltage required to emulate a variable reactance can be injected into the AC system by means of synthetic sources, which can be realized using static power converters.

另一方面，模拟可变电抗所需的电流或电压可以通过合成电源注入交流系统。合成电源可以利用静态功率变换器实现。

13. The unified power flow controller (UPFC) was proposed for real-time control and dynamic compensation of AC transmission systems, providing the necessary functional flexibility required to solve many of the problems facing the utility industry.

统一潮流控制器（UPFC）可用来对交流输电系统进行实时控制和动态补偿，这为解决许多电力工业面临的问题提供了必要的灵活性。

14. It was found later that the bias power supply is unnecessary, the bias current could be provided by the source by proper thyristor control.

后来发现采用合适的晶闸管控制方法可以不使用偏置电源，并且所需的偏置电流可以由电源提供。

Chapter 7
Renewable Energy and Distributed Generation

7.1 Introduction

Now we are facing two major energy challenges. The first is sustainability as worlds CO_2 emissions are forecast to rise. The second is security of supply of the word is becoming more dependent on imported fuels. Today these account for 50% of our energy consumption, but the 2030 figure is forecast to be around 65%. The Council of the European Union recently agreed that Europe should develop a sustainable and integrated climate and energy policy. Building on Denmark's traditionally strong environmental profile, the Danish government earlier in 2007 put forward the document "A Visionary Danish Energy Policy for the Period up to 2025". This aims to stabilize energy consumption at its current level, and calls for a considerable increase in the use of renewable energy.

In its fourth assessment report of the Intergovernmental Panel on Climate Change (IPCC) says that if we want to stabilize CO_2 at the low level-around 500 ppm—needed to limit the global average temperature rise to 2.5-3.0℃, CO_2 emissions must peak soon and then decline. The IPCC states that we must take action now if we are to stabilize CO_2 at a low level. With the global expansion of intermittent renewable energy technologies comes the pressing need to solve the problem of long-term variability.

Carbon capture and storage (CCS) has moved to centre stage in the last few years as a serious option for large scale CO_2 emissions mitigation.

Wind energy has seen an average annual world market growth of 17% over the last five years in terms of installed capacity. European countries are leaders in the deployment of wind energy: half of all the new wind turbines installed in 2006 were in Europe.

Fuel cells are within the next five years at the entrance to their break-through. They will be used in three main applications: stationary power generation, transport, and portable equipment.

Solar cells (PV) represent one of the fastest-growing renewable energy technologies, with a global annual growth of more than 40%. Polymer solar cells are a promising new technology. The falling cost of PV systems will eventually make PV electricity competitive in Denmark.

Bioethanol is promising as a transport fuel. The best alternative is second-generation bioethanol from waste materials such as straw. Other liquid transport fuels are biodiesel, synthetic gasoline and diesel produced from gasified biomass. Biomass can also be used for heating, replacing oil or natural gas that can be used as motor fuel.

Coal has, as the most abundant fossil fuel, gained renewed interest. Most of Denmark's electricity comes from the combustion of pulverized coal, and Danish coal-fired power plants lead in the world in energy efficiency. Nevertheless, coal will only be an option for the future if we can cost-effectively reduce CO_2 emissions from coal combustion. This can be done in three ways: increase the energy conversion efficiency; switch to a fuel with a lower fossil carbon content (including biomass); and capture and store CO_2 produced during combustion.

Nuclear fission is a major source of carbon-free energy. It provides 15% of the world's electricity. 15 countries are currently building new nuclear power stations, and a further 25 plan to do so. In contrast to previous prognoses, now that nuclear power is assumed to increase by 15% by 2030.

Geothermal energy has been shown to have a huge potential. With the present high oil prices, the number of towns embarking on geothermal projects is increasing.

Wave power has gained renewed interest in the world. Examples are Wave Dragon and Wave Star. These demonstration projects are very successful as a starting point for the commercial development of this technology.

In promotion of the renewable energy policy, the large scale expansion of constructions of the wind and solar power is developing in China, whose random high-power electric energy fluctuation causes huge shock to the power grid. Confronting this important problem of the power system safe operation, it is really top urgent to build a feasible solution to develop the large-scale renewable energy.

Distributed generation system (DGS) based on various renewable energy resources is the important approach to the development of clean energy, improving the reliability of power supply and enlargement of the power system capacity. Compared to the traditional centralized power supply system, DGS has many advantages: easy start-stop, good peak shaving, beneficial to load balance and less investment, yield faster result, and satisfying power supply demand in the special occasion and less transmission loss, improving disaster level.

7.2 Wind Power

Wind energy is a fast-growing interdisciplinary field that encompasses multiple branches of engineering and science. According to the World Wind Energy Association, the global installed capacity of wind turbines grew at an average rate of 27% per year over the years 2005-2009. At the end of 2009, the installed capacity in the United States was about 35000 MW, while the worldwide installed capacity was approximately 160000 MW. Wind is recognized worldwide as a cost effective, environmentally friendly solution to energy shortages. Although the United States receives only about 2% of its electrical energy from wind, that figure in Denmark is approximately 20%. The comprehensive report by the U. S. Department of Energy lays the framework for achieving 20% of the

U. S. electrical energy generation from wind by the year 2030. This report covers technological, manufacturing, transmission and integration, market, environmental, and sitting factors.

Despite the growth in the installed capacity of wind turbines in recent years, engineering and science challenges remain. Because larger wind turbines have energy-capture and economic advantages, the typical size of utility-scale wind turbines has grown by two orders of magnitude over the last three decades. Since modern wind turbines are large, flexible structures operating in uncertain environments, advanced control technology can improve their performance. For example, advanced controllers can help decrease the cost of wind energy by increasing turbine efficiency, and thus energy capture, and by reducing structural loading, which increases the lifetimes of the components and structures. The goal of this section is to describe the technical challenges in the wind industry relating to control engineering.

Although a wind turbine can be built in either a vertical-axis or horizontal-axis configuration, we focus on horizontal-axis wind turbines (HAWTs) because they dominate the utility-scale wind turbine market. At the utility scale, HAWTs have aerodynamic and practical advantages. Smaller vertical-axis wind turbines (VAWTs) are more likely to use passive rather than active control strategies. The generating capacity of commercially available HAWTs ranges from less than 1 kW to several megawatts. Active control is more cost effective on larger wind turbines than smaller ones, and therefore this section focuses on HAWTs whose capacity is 600 kW or larger.

The next section describes the configurations and basic operation of wind turbines. We then explain the layout of a wind turbine control system by taking a "walk" around the wind turbine control loop, with discussions on wind inflow characteristics, available sensors and actuators, and turbine modeling for use in control. Subsequently, we describe the current state of wind turbine control, which is followed by a discussion of the issues and opportunities in wind turbine and wind farm control. At the end, we give some concluding remarks.

7.2.1 Wind Turbine Basics

The main components of a HAWT that are visible from the ground are the tower, nacelle, and rotor, as shown in Fig. 7.1. The wind first encounters the rotor on this upwind horizontal-axis turbine, causing it to spin. The low-speed shaft transfers energy to the gearbox, which steps up in speed and spins the high-speed shaft. The high-speed shaft causes the generator to spin, producing electricity. Also shown is the yaw-actuation mechanism, which is used to turn the nacelle so that the rotor faces into the wind. The airfoil-shaped blades capture the kinetic energy of the wind and transform it into the rotational kinetic energy of the wind turbine's rotor. The rotor drives the low-speed shaft, which in turn drives the gearbox. The gearbox steps up the rotational speed and drives the generator by means of the high-speed shaft. The gearbox, high speed shaft, and generator are housed in the nacelle, along with part of the low-speed shaft. Direct drive configurations without gearboxes are being developed to eliminate costly gearbox failures.

Wind turbines may be variable or fixed speed. Variable speed turbines tend to operate closer to their maximum aerodynamic efficiency for a higher fraction of the time but require electrical power

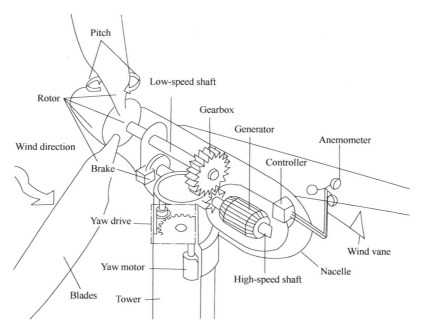

Fig. 7.1 Wind turbine components

processing so that the generated electricity can be fed into the electrical grid at the proper frequency. Variable-speed turbines are more cost effective and thus more popular than constant-speed turbines at the utility scale because of improvements in generator and power electronics technologies. Variable-speed operation can also reduce turbine loads, since sudden increases in wind energy due to gusts can be absorbed by an increase in rotor speed rather than by component bending.

The goals and strategies of wind turbine control are affected by the turbine configuration. A HAWT can be upwind, with the rotor on the upwind side of the tower, or downwind, with the rotor on the downwind side of the tower. This choice affects the turbine dynamics and thus the structural design. A wind turbine can also be variable pitch or fixed pitch, meaning that the blades may or may not be able to rotate about their longitudinal axes. Variable-pitch turbines might allow all or part of their blades to rotate along the pitch axis. Fixed-pitch machines are less expensive to build, but the ability of variable-pitch turbines to mitigate loads and affect the aerodynamic torque has driven their dominance in modern utility-scale turbine markets. The example given in Fig. 7.2 shows power curves for a 2.5 MW variable-speed turbine and a 2.5 MW fixed-speed turbine, as well as a curve showing the available wind power for a turbine with the same rotor size as these two turbines.

For both turbines, when the wind speed is low, the power available in the wind is low compared to losses in the turbine system; hence, the turbines are not run. When the wind speed is above the rated wind speed, power is limited for both turbines to avoid exceeding safe electrical and mechanical load limits. In this example, low wind speed is considered to be below 6 m/s, whereas high wind speed is above the rated wind speed of 11.7 m/s.

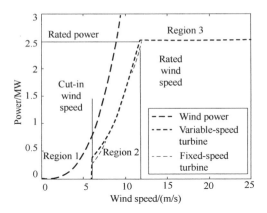

Fig. 7.2　Illustrative power curves

The main difference in the example shown in Fig. 7.2 between the two types of turbines appears for mid-range wind speeds. The wind power curve shows the power available in the wind for the example variable-and fixed-speed turbines with the same rotor diameter. The turbines are started up at 6 m/s wind speed when there is enough wind to overcome losses and produce power. Above 11.7 m/s wind speed, power is limited to protect the turbines' electrical and mechanical components. Both turbines generate the same power at the fixed speed turbine's 10 m/s design point, but the variable speed turbine generates more power at all wind speeds in Region 2, which encompasses wind speeds between 6 and 11.7 m/s, with a maximum difference of about 150 kW at 6 m/s. Except for the fixed speed turbine's design operating point of 10 m/s, the variable-speed turbine in Fig. 7.2 converts more power at each wind speed than the fixed-speed turbine. This example illustrates the fact that variable-speed turbines can operate at maximum aerodynamic efficiency over a wider range of wind speeds than fixed-speed turbines.

The wind speed probability distribution can be modeled as a Weibull function, with scale and shape parameters that define the function. In the case of a Weibull distribution having a shape parameter $k = 2$ and scale parameter $c = 8.5$, the variable-speed turbine captures 2.3% more energy per year than the constant speed turbine. A wind farm rated at 100 MW and operating with a 35% capacity factor can produce about 307 GW·h of energy in a given year. If the cost of energy is US $0.04 per kW·h, each GW·h is worth about US $40000, and each 1% loss of energy on this wind farm is equivalent to a loss of US $123000 per year.

Not shown in Fig. 7.2 is the high wind cut-out, the wind speed above which the turbine is powered down and stopped to avoid excessive operating loads. High wind cut-out typically occurs at wind speeds between 20 m/s and 30 m/s for utility-scale turbines.

Momentum theory using an actuator disc model of a wind turbine rotor shows that the maximum aerodynamic efficiency, called the Betz limit, is approximately 59% of the wind power. The aerodynamic efficiency, which is the ratio of the turbine power to the wind power, is given by the power coefficient

$$C_p = \frac{P}{P_{wind}} \tag{7-1}$$

where P is the instantaneous turbine power and

$$P_{wind} = \frac{1}{2}\rho A v^3 \tag{7-2}$$

is the instantaneous power available in the wind for a turbine of that rotor diameter. In Equation (7-2), ρ is the air density, A is the swept area πR^2 of the rotor, R is the rotor radius, and v is the instantaneous wind speed, which is assumed to be uniform across the rotor swept area. The swept area is the area of the disk circumscribed by the blade tip.

Finally, utility-scale wind turbines are either two or three bladed. Two-bladed turbines typically use a teetering hinge to allow the rotor to respond to differential loads. This teeter hinge allows one blade to move upwind while the other moves downwind in response to differential wind loads, much like a seesaw allows one child to move up while another moves down. For a turbine with an even number of blades placed symmetrically around the rotor, when one blade is at the uppermost position, another blade is in the slower wind caused by either the tower shadow behind the tower or the bow wake in front of the tower. This discrepancy is exacerbated by typical wind shear conditions, which result in higher wind speeds higher above the ground. Three-bladed turbines tend to experience more symmetrical loading than two-bladed turbines, but at a 50% increase in blade cost.

7.2.2 A Walk Around the Wind Turbine Control Loops

In designing controllers for wind turbines, it is often assumed as in Equation (7-2) that the wind speed is uniform across the rotor plane. However, as shown by the instantaneous wind field in Fig. 7.3, the wind input may vary in space and time over the rotor plane. Since the wind speed varies across the rotor plane, wind speed point measurements convey only a small part of the information about the wind inflow. Rotor speed is the only measurement required for the baseline generator torque and blade-pitch controllers described in this sector. The deviations of the wind speed from the nominal wind speed across the rotor plane can be considered disturbances for control design.

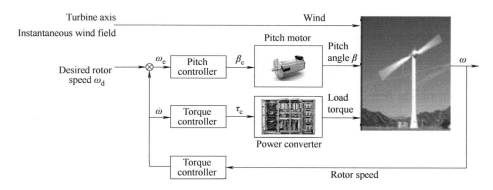

Fig. 7.3 Wind turbine control feedback loops

Utility-scale wind turbines have several levels of control, namely, supervisory, operational, and subsystem. As shown in Fig. 7.4, the top-level supervisory control determines when the turbine starts and stops in response to changes in the wind speed and also monitors the health of the turbine. The operational control determines how the turbine achieves its control objectives. The subsystem controllers are responsible for the generator, power electronics, yaw drive, pitch drive, and remaining actuators. In this section, we move through the operational control loops shown in Fig. 7.3, describing the wind inflow, sensors, actuators, and turbine model while treating the subsystem controllers as black boxes. The pitch and torque controllers in Fig. 7.3 are discussed in the section "Feedback Control". The details of the subsystem controllers are beyond the scope of this sector.

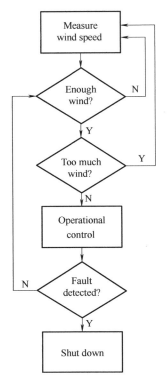

Fig. 7.4 Wind turbine supervisory control logic

1. Wind Inflow

The differential heating of the Earth's atmosphere is the driving mechanism for the wind. Various atmospheric phenomena, such as the nocturnal low-level jet, sea breezes, frontal passages, and mountain and valley flows, affect the wind inflow across a wind turbine's rotor plane, which spans from 60 m to 180 m above the ground for megawatt utility-scale wind turbines. Given the large rotor plane and the variability of the wind, hundreds of sensors would be required to characterize the spatial variation of the wind speed encountering the entire span of each blade. The available wind resource can be characterized by the spatial or temporal average of the wind speed; the frequency distribution of wind speeds; the temporal and spatial variation in wind speed; the most frequent wind direction, also known as the prevailing wind direction; and the frequency of the remaining wind directions. The probability of the wind speed being above a given turbine's rated wind speed can be used to predict how often the turbine operates at its maximum, that is, rated, power generation capacity. The capacity factor CF is defined by the ratio

$$CF = \frac{E_{out}}{E_{cap}}$$

Where E_{out} is a wind turbine's energy output over a period of time and E_{cap} is the energy the turbine would have produced if it had run at rated power for the same amount of time. Capacity factor can also describe the fraction of available energy captured by N turbines in a wind farm.

To predict the capacity factor and maintenance requirements for a wind turbine, it is useful to understand wind characteristics over both long and short time scales, ranging from multiyear to subsecond. Determining whether a location is suitable and economically advantageous for sitting a wind turbine depends on the ability to measure and predict the available wind resource at that site. Significant variations in seasonal average wind speeds affect a local area's available wind resource over the course of each year. Wind speed and direction variations caused by the differential heating of the Earth's surface during the daily solar radiation cycle occur on a diurnal, that is, daily time scale. The ability to predict hourly wind speed variations can help utilities to plan their energy resource portfolio mix of wind energy and additional sources of energy. Finally, knowledge of shortterm wind speed variations, such as gusts and turbulence, is used in both turbine and control design processes so that structural loading can be mitigated during these events. Since wind inflow characteristics vary temporally and spatially across the turbine's rotor plane, assuming uniform constant wind across the rotor plane is problematic for control design for large rotor sizes. The uniform wind assumption, which is used in Equations (7-1) and (7-2), can lead to poor predictions of the available wind power and loading on the turbine. Especially problematic are nonuniform winds such as low-level jets. Analysis indicates that rotor-sized or smaller turbulent structures in the wind can cause more damage than turbulent structures that are larger than the rotor.

Improved capabilities for measuring and predicting turbulent events are needed, and this area of research is active among atmospheric scientists.

2. **Sensors**

As shown in Fig. 7.3, the rotor speed measurement is used in feedback for basic control. Since the gearbox ratio is known, the rotor speed can be measured on either the high-speed or low-speed shaft. Rotor speed measurements can be used for speed and power control but might not be suitable for more sophisticated control objectives, such as reducing torsional oscillations in the drive train.

In addition to rotor-speed measurements, anemometers are used for supervisory control, in particular to determine whether the wind speed is sufficient to start turbine operation. Fig. 7.5 shows sonic and propeller anemometers on a meteorological tower. For measuring wind speed and wind direction, the majority of turbines have an anemometer and wind vane located on top of the nacelle at approximately the hub height. Because of the interaction between the rotor and the wind, the measurements are distorted in both upwind and downwind turbines.

Fig. 7.5 The rotor speed measure set

Power measurement devices are used to assess energy generation. Additional sensors that are sometimes found on wind turbines include strain gages on the tower and blades, accelerometers, position encoders on the drive shaft and blade-pitch actuation systems, and torque transducers.

Sensors used for control, especially rotor or generator speed sensors and wind vanes, must be reliable. Fault detection can be used to identify faulty sensors and practical problems related to the size of turbines and their noisy environments can hinder sensor failure diagnoses. Calibration drift is a typical failure mode, especially for strain gages and accelerometers, and thus controllers that depend on sensors prone to drift must be robust to calibration errors such as unknown biases.

3. Actuators

Utility-scale wind turbines typically have up to three main types of actuators. A yaw motor, which aligns the nacelle with the wind, provides active yaw control actuation for large turbines. However, due to dangerous gyroscopic forces, HAWTs must not be yawed at high rates. Consequently, the maximum yaw rate of a large turbine is typically limited to less than 1°/s. Thus, investigation of advanced controllers for yaw control provides less potential for benefit compared to advanced controllers for the remaining actuators. The second type of actuator is the generator which, depending on the generator and power processing equipment, can be commanded to follow a desired torque or load. Generator torque control, performed using the power electronics, determines how much torque is extracted from the turbine. The generator and power electronics produce a load torque based on the separation of magnets in the generator's stator and rotor, which should not be confused with the turbine's rotor. The type of magnets and methods for producing the separation depend on the type of generator and power electronics. Although the net torque on the rotor depends on the input torque from the wind and the load torque from the generator, the generator torque can be used to affect the acceleration and deceleration of the rotor. The fast time constant of the generator torque, which is at least an order of magnitude faster than that of the rotor speed, makes generator torque an effective actuator for controlling rotor speed. The third type of actuator is the blade-pitch motor.

7.2.3 Modeling and Control of Wind Farms

Wind turbines are often located with other turbines in wind farms to reduce costs by taking advantage of economies of scale. Fig. 7.6 shows turbines in a linear array. Turbines on wind farms can be located along a single line, in multiple lines, in clusters, in grids, or in configurations based on geographical features, prevailing wind direction, access requirements, environmental effects, safety, prior and future land use including farmland and ranchland, and visual impact.

From a control systems perspective, wind farm research is focused mainly on either control of the electricity generated by the turbines or coordinated control of the energy captured by individual turbines to minimize the negative effect of aerodynamic interaction. Although various types of generators can be used, doubly-fed induction generators are increasingly used in wind turbines. Standards governing the wind farm's interconnection to the utility grid vary by location, but typically it is expected that wind farms are equipped with strategies for voltage control, do not contribute to

Fig. 7.6 Turbines in offshore wind farms

grid faults, and are not damaged by grid faults. Voltage stability and the uninterrupted operation of a wind farm connected to an electric grid during a grid fault is an active area of research, as is the use of wind turbines to protect against grid faults. Individual turbines and older farms with comparatively small capacity had little effect on the grid.

Control of active and reactive power supplied by wind farms to the utility grid is also a major area of research. Experiences at Denmark's Horns Rev, the first wind farm equipped with advanced control of both active and reactive power, can provide guidelines for newer wind farms. Lyapunov-based strategies can be used to damp the network electromechanical oscillations of a wind farm's power output while trying to minimize the number of sensors required in the farm. Electrical system control over multiple levels is typically used to achieve multiple objectives on a wind farm and relies on models of wind farms from an electrical perspective.

The aerodynamic interaction of turbines on a wind farm is not as well understood as the electrical interconnection of the turbines. While wind farms help to reduce the average cost of energy compared to widely dispersed turbines due to economies of scale, aerodynamic interaction among turbines can decrease the total energy converted to electricity compared to the same number of isolated turbines operating under the same wind inflow conditions. Turbines on a wind farm are typically spaced farther apart in the direction parallel to the prevailing wind direction, known as downwind spacing, than in the perpendicular direction, known as crosswind spacing. Downwind spacing is often eight to ten rotor diameters, and crosswind spacing is often four to five rotor diameters, although the exact distances chosen vary with geography and additional factors. Shorter cross wind spacing distances can reduce land cost and are beneficial at locations where the frequency distribution of the wind direction is skewed heavily toward the prevailing direction.

The array efficiency η_A of a wind farm is given by

$$\eta_A = \frac{E_A}{E_T N} \tag{7-3}$$

Where E_A is the annual energy of the array, E_T is the annual energy of one isolated turbine, and N is the number of turbines in the wind farm. Array efficiency greater than 90% can be achieved when downwind spacing of eight to ten rotor diameters and crosswind spacing of five rotor diameters are used.

Since wind turbines can slow the wind over a distance of 5-20 km, turbines arranged in a wind farm interact aerodynamically. Coordinated control of the turbines can reduce the negative effects caused by this aerodynamic interaction. It can be shown that having each wind turbine in an array extract as much energy as possible does not lead to maximal total overall energy capture across the entire array because the turbines on the upwind side of the farm extract too much energy, slowing the wind too much before it reaches the remaining turbines. The spatial variation of turbines on a wind farm often results in power smoothing, where the standard deviation of the power produced by multiple turbines is less than the standard deviation of the power produced by each individual turbine. This effect is caused by different wind gusts and lulls hitting different turbines on the farm at different times.

7.3 Solar Power

7.3.1 Solar Cells

Solar cells are devices which absorb light and convert it into electrical energy. The absorption of light in semiconductors creates additional electrical charge carriers, both electrons and holes equally. If an electric field exists within the semiconductor, the negative electrons and positive holes move in opposite directions, and this electrical charge separation results in the creation of a voltage. This is the photovoltaic effect—the creation of a voltage by the action of light. The movement of the electric charges constitutes an electric current, so both current and voltage are generated simultaneously, as shown schematically in Fig. 7.7.

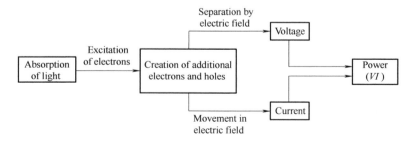

Fig. 7.7 Schematic representation of the photovoltaic effect

The simplest way to establish an electric field in a semiconductor is to make a P-N junction. The electric field at the junction attracts electrons from the P-side and forces them to the N-side making it negatively charged. Similarly holes from the N-side are forced to the P-side, making this positively charged, and thus creating a voltage. The electric current is fed via the metal contacts on the P and then N layers to an external load.

The requirements for a solar cell are thus:

1) A semiconductor diode of large area, to collect as much light as possible.

2) Electrical contacts on each side of the diode. One side must collect the incident light, so the contact must be as transparent as possible. It can be a transparent conductor or fine metal fingers spaced widely apart, as shown in Fig. 7.8.

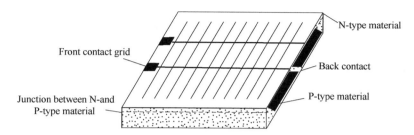

Fig. 7.8   Structure of solar cell

The first criterion for choosing a solar cell material is its energy gap. The voltage which a cell can generate is a fraction (known as the voltage factor) of its energy gap, so the larger the energy gap the higher the voltage. A semiconductor will absorb light only if the energy of the photons is greater than the energy gap of the semiconductor. For a light source such as the sun with its energy spread over a wide range of photon energy (colours) the higher the energy gap the smaller the fraction of sunlight which can be absorbed, and therefore the lower the current. There is thus an optimum energy gap for a solar cell at which the product of voltage and current is a maximum. The optimum value is about 1.4 eV, but values from 1 eV to 1.6 eV will provide efficient conversion. There are a number of well-known semiconductors close to the optimum energy gap, including silicon, gallium arsenide, indium phosphide and cadmium telluride. Other materials close to the optimum include copper indium diselenide and amorphous silicon and alloys of these materials.

The efficiency of a solar cell is determined by three physical factors and three main engineering factors. Sunlight whose energy is less than the energy gap is not absorbed and so not converted into electricity. Sunlight whose energy is higher than the energy gap can donate only energy equal to the energy gap and the rest is dissipated. Finally, the voltage output is only two-thirds or so of the energy gap, with again a dissipation of the excess energy. These three factors determine the maximum possible efficiency. The three engineering factors of reflectance, series resistance and recombination reduce the efficiency of actual cells below the maximum.

Since the sunlight has at most a power density of 1000 W/m^2, solar cells must cover a large area to collect substantial amounts of power. This large area must be produced and deployed at as low a cost as possible and the cells should ideally be made by a low-cost mass-production process. The ideal material should be cheap, abundant and environmentally benign and suffer no degradation over 30 years or so of operation. Such an ideal material has not yet been found, but the materials in commercial use, or in commercial development, are well able to provide the basis of an expanding and effective industry for the next 20-30 years. Beyond that time one or more of the novel materials presently being studied in research laboratories will form the next generation and carry the industry forward.

7.3.2 Integrated Solar Home System

To date, many traditional solar home systems (SHSs) have consisted of separate components which required assembly by trained individuals and were also more susceptible to failure and maintenance. As a result, many SHSs in remote areas have not fulfilled their desired lifecycles or simply have not functioned at all. Thankfully, a solution to these problems has arrived—the newly developed integrated solar home system (I-SHS). Within this new system all components such as the support structure, foundation, PV modules, charge controller, DC-AC converter and wiring are re-assembled by the manufacturer. Benefits of the new system are ease of assembly and maintenance combined with an associated reduction in cost and failure—critical aspects to consider for remote and impoverished regions.

1. **Composition of the System**

For example, there are more than 20 million people in Brazil (approx. 42% of the rural population) do not have access to electricity. Grid line extension is a rather costly option because distances are long and average consumption is low. According to Messenger and Ventre installation of each kilometer of a simple 115 V line extension costs in the vicinity of 50000€—depending on the region and the environment. In addition, costs for surveillance and maintenance are considerable due to exposure of grid lines to hostile conditions (e.g. vandalism, thunderstorms, vegetation, flooding).

Conversely, a means to supply remote areas with electrical energy is SHS. Apart from ecological advantages, in many cases this option is also the most economic way to electrify rural areas—especially when consumption is low and grid extension is long. But even this most economic option often has a price that is too high for the wide spread use of traditional SHS. According to a recent report by Kister, about 60% of the traditional SHS installed in Brazil by the PRODEEM program (rural electrification by PV) are no longer working or never worked at all. Reasons include a shortage of trained staff which resulted in inadequate installations and improperly maintained systems. Another reason for the high failure rate of the systems was on the load side because traditional SHSs were installed with DC outputs only (which have the advantage that the user is not able to connect to common and typically inefficient electro-domestic equipment such as filament light bulbs). On the other hand, it is rather difficult to find replacements for such supplies in retail outlets. Furthermore, DC equipment (e.g. fluorescent lamps) is more expensive than AC equipment. System components could be manipulated by the user in an undesired manner (e.g. bypassing of the charge controller). The newly developed I-SHS minimizes or eliminates these problems.

Fig. 7.9 shows the basic layout of the system: The PV generator consists of two parallel-connected frameless 30 WP modules. Located in the foundation structure are a maintenance-free lead acid battery (12 V, 105 Ah) and a 200 W sine wave inverter (115 V, 60 Hz) with an integrated charge controller (6 A). A water tank cools all components. The output leads to a regular AC plug. Additionally, five Pt 100 temperature sensors have been integrated into the prototype. Through computer based measurements—utilizing a 16-channel-USB-DAC card and LabView software—the temperature profile has been improved through design changes to achieve optimal cooling in a future prototype. Ambient temperature, irradiance and the electrical parameters (V, I, P) are also monitored (see Fig. 7.9 and Fig. 7.10). All components are contained in a waterproof epoxy fiberglass tank. The prototype is 1.37 m long, 0.76 m high and 0.5 m deep and has a volume of 0.3 m^3. A module elevation angle of 30° was chosen to achieve a good yield even in winter in most parts of Brazil. The tank has a volume of almost 300 L, which results in a weight of at least 300 kg, when full.

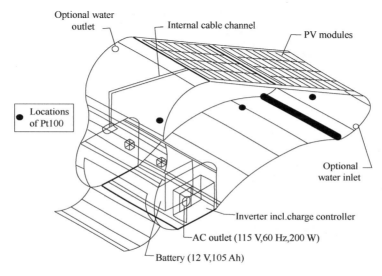

Fig. 7.9 Structure and components of the I-SHS

2. Temperature Dependence

The electrical power generation of a solar cell depends on its operation temperature. While the short circuit current (I_{sc}) increases slightly with increasing temperature, the open circuit voltage (V_{oc}) decreases significantly (about −2.3 mV for each K) with increasing temperature, leading to an electrical yield reduction of −0.4% K^{-1} to −0.5% K^{-1} for mono-and multi-crystalline silicon solar cells which are used in most SHS applications. Fig. 7.11 shows the I-V characteristics for a typical multi-crystalline silicon solar cell at different temperatures together with the operation points for maximum power generation.

3. Temperature Reduction

While efficiency and electrical yield is decreasing with increasing operation temperature, the idea to keep the system at low temperatures is quite evident. The energy consumption of an active cooling system would not be compensated by the gain in increased energy generation, at least for small systems. Operation temperatures were kept at low levels by mounting the module on a water

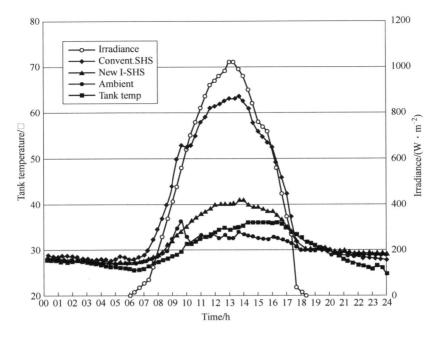

Fig. 7.10 Temperature measurements of the I-SHS during a clear day (irradiance): The lower module temperature (new I-SHS) and water temperature in upper part of the container (tank temp.), in comparison to a conventional SHS and ambient temperature

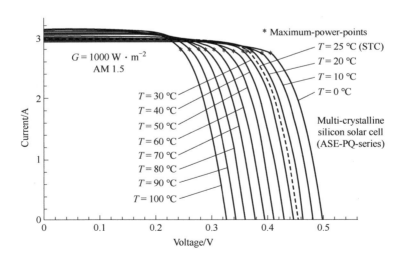

Fig. 7.11 *I-V* characteristics at different temperatures of a typical multi-crystalline silicon solar cell

filled tank. This allows for an effective reduction of operating cell temperatures without spending any energy for refrigeration. The water virtually soaks up the heat flow generated by the module. Due to the high thermal capacity of the incorporated water, the temperature increases gradually. The principle was proven and validated with different prototypes in Europe and in Africa built over previous years.

4. History

The first cooling device which followed the 'cooling by an extended heat capacity' concept was built in 1992. The tank was integrated into the original framing of a M55 PV module by SSI (former Siemens Solar Industries, now Shell Solar) with a volume of 12 L, so it could be used with conventional mounting. This prototype provided a 2.6% increase in daily electrical energy yield. Subsequent tests, which utilized latent heat storage material (sodium sulphate), showed significantly better results but caused severe corrosion.

The second prototype built in 1994 had a much larger water tank which served also as the module's foundation, stand and mounting structure (TEPVIS—Thermal Enhanced PV module with Integrated Standing). It was tested with a M55 in Berlin which showed an energy gain of up to 12%, and with PQ 10/40 devices in Bulawayo, Zimbabwe which proved an increase of 9.5%. The gain in Zimbabwe was lower due to reduced water circulation and more stratification (the upper part of the tank got considerably warmer than the lower one). The inclination of the module plane in Zimbabwe was much lower, according to the latitude of Bulawayo, thus reducing the thermo-siphon effect. The device in Berlin was also equipped with an additional plate inside the tank, in parallel to the module at a distance of 12 cm, forming a kind of chimney and thus enhancing circulation. Nevertheless, the PV conversion efficiency of the cooled module was considerably above the reference module during at least 95% of the day.

To eliminate possible measurement errors which may have been caused by the differing electrical properties of the systems, all modules (reference and cooled ones) were interchanged and retested. The modules had been operated continuously at the maximum power point (MPP) by manual tracking of an ohmic load together with power metering. The application for the new I-SHS was carried during the first quarter of 2002.

Once placed at an appropriate site the I-SHS is immediately ready to supply small AC loads (illumination, air-fan, radio etc.). Additionally it is capable of supplying the hot water needed for a small household. Several systems can be combined to fulfill higher power needs without a redesign of the system. Without having higher costs than conventional SHSs, and featuring favorable balance of system (BOS) and the generation of more energy, the I-SHS is an efficient means to successfully electrify remote areas.

7.3.3 PV Utilization in China

The pioneer application of PV in China is utilized to the exploitation of space domain by Chinese scientist due to the price of PV is very costly, such as which is used to the power supply of secondary planet and spacecraft, and the weight of PV has a share of 10%-20% of the whole secondary planet weight, and the price of PV has a share of 10%-30% of the whole secondary planet price. So the weight and price of PV must be decreased in order to decrease the launch price of secondary planet, which is the research direction of space application in future. PV has been used in various domains with the development of economy and society at present. Sea sail assistance is very important to instruct the sail direction of ship. If the load of system is enough small, the efficiency of system is enough high, which can be provided the power supply by the PV, and the cost is decreasing with the technology progress. Now the PV system has been used to the wireless communications in remote zone and carry-home PV power supply and cathode protection and commodity. The most broad application practices of PV in China are described in the section.

1. The City Road Lighting System

It is well known that hundreds of big cities lie in the large soil of China, and there are more than 10 million street lamps in those cities. The annual total amount of whole city road lighting system is more than one billion kW·h, and billions of dollars is expended in order to pay the expense of electric power, which had become a enormous burden of Chinese government. The abundant primary resource is consumed at the time, and the wastage of coal is more than two million tons, and amass of CO_2, SO_2 and NO_X are vented, and amass of castoff pollute the environment and water and air. Fortunately, the problems have been considered by the Chinese central government and local government and ordinary people. The solar energy is utilized in the city road lighting system by some local governments in order to improve the local environment, i.e., solar energy street lamp, solar energy community lighting and solar energy scenery lighting. The solar energy street lamp has better competition and is more popular. More and more cities in China begin to replace the conventional street lamp by using the solar energy street lamp. For instance, there are more than 3000 solar street lamps by using the city lighting system in Binzhou. The whole street lighting system is replaced by using the solar street lamp in Linan, Zhejiang province. Moreover, the annual electric power cost of conventional street lamp in Hangzhou is 0.3 billion RMB. It is estimated that the investment of solar LED street lamp is equal to the conventional lighting system during 3 years. The great economy income is received from the renewable solar street lamp during the remaining years. Synchronously, the enormous income of environment is gained.

Other solar energy lighting systems have been used to improve the life of common people, such as court lighting, lawn lighting and scenery lighting. The solar lighting systems improve the habitation and life quality of citizen, which brings the enormous environment and economy benefit. For example, the solar street lamp is used to improve the lighting condition of Shitai freeway in 2006, and the total investment is more than three million RMB. The total of traffic flux is increasing 13% compared with the corresponding period of last year, and the annual total of traffic accident in winter is decreasing from 50 to 25, and the economy loss is decreasing from 2 million RMB to 12000 RMB, and the total of injured people is decreasing from 41 to 25 at the same time. Moreover, the scenery lighting is used to improve the sight of hilly country park in Xiamen, and the total of solar scenery lamp is more than 200. The lighting system of remote village in Yangzhong is achieved by using the solar street lamp. As mentioned above, the requirement of solar street lamp is enormous with the sustainable development of China in future.

2. Solar Water Pump

The economy and zoology is very brittle in west zone of China. And the area of west zone is more than 4 million km^2, and the population is more than 0.2 billion, and the zone contains abundant natural resource. Unluckily, the water is exceeding lack in the northwest zone of China, and the desert zone of northwest China in 2007 is more than 1.3 million km². Based on the data of Chinese Forest Bureau, the annual increscent desert area in 2001 is more than 2000 km². For instance, Xinjiang province lies in the northwestward of China, the area is more than 1.6 million km², and the natural resource is abundant. But the Takelamagan desert lies in the middle part of Xinjiang province, and the area is more than 0.33 million km², and Xinjiang is divided into two parts. Fortunately, Chinese central government and local government have realized the problem of increasing desert area, and some actual actions have been implemented to improve the environment and zoology of northwest zone. And the groundwater of desert zone is abundant in Chinese northwest zone, which gives a hope to banish the desert. Some electric power is used to pump groundwater by using water pump. As mentioned above, the environment and zoology is very brittle, and the desert zone is far away from main power lines, and the traffic is not convenient. Simultaneously, the high solar irradiation exists in the large northwest zone. The solar water pumping has great potential to banish the desert and improve the irrigation area of northwest farmland. Some actual applications have been used to improve the the zoology of northwest and traffic, such as the desert road, protect desert oasis and solidify desert. For instance, the Talimu desert road is laid to link the north and the south of Xinjiang province in 1995, and the length of which is more than 550 km. At present, more than 5 desert roads have been laid in the last 20 years.

In an example of solar water pump application in remote villages of northwest China, the area of grassland in China is more than 0.4 billion hectares. The area is 0.102 billion hectares, which is irrigated by local people. However, the environment and zoology of Chinese northwest are brittle due to the annual rainfall is less than evaporation. The area of desert is increasing in the last 50 years, and the total of desert area in 2007 is more than 1.3 million km². The water pump for irrigation is used to improve the local environment by local people. It is well known that the economy of Chinese northwest is very poor, and the local people cannot support the costly fee of diesel water pump, and the diesel oil will destroy the brittle zoology of northwest. In other words, solar water pump gives a hope to the people to improve the local zoology, but the solar water pump has a higher price than the diesel water pump. Fortunately, the central government and local government give some assistance to increase the popularization of solar water pump. An actual solar water pump system is described in the section, which lies in Neimenggu province. The end user' name is Nashun, and the grassland area of family is 5.3 ha. The system contains: a well, a PV, a water pump, a converter/inverter and some sprinklers. The capacities of PV and water pump are 3 kW and 500 W/24 V, respectively. The total of investment in 1999 is more than 60000 dollars. The total of livestock is more than 370 by using the solar water pump. In China, the irrigation area of solar water pump in 2003 is 534 ha. The object area in 2010 is more than 392000 ha, and the need of PV is more than 261 MW. As mentioned above, the prospect of solar water pump in China is great in future.

3. **Distributed Generation (DG)**

The large-scale distribution network (LSDN) is considered by Chinese government in past 30 years, and the accumulative total amount of electric energy in 2007 is more than 0.713 billion kW. However, a potential danger exists in the LSDN because the modern people are more and more dependent on the electric power supply. If an electric network occurs an accident, which will affect the daily life of millions of people, and the unpredictable accident will stop the factory production and the society movement because the electric power is cut. For an instance, the northeast of USA and the east of Canada are cut the electric power by an unpredictable electric network accident in 2003 and more than 50 million people are affected during the power cut, and the daily economy loss is more than 30 billion dollars. So a credible power supply must be found in order to conquer the unforeseen accident. Fortunately, the solar energy is not big affected in the natural disaster and accident, and a solar distributed generation can partially afford the electric supply. With the improvement of people life, more and more people and Chinese government have realized the importance of DG to improve the security of electric power supply. For instance, millions of cattle farmers working in the widest northwest zone of China, the herd and cattle farmer will move with various seasons. Because the browse zone is far away from main power lines, so they can conveniently gain the electric power by using the small DG units. In a word, the DG is important to improve the security of electric power supply and the life quality of common people.

Some actual applications have improved the life of ordinary people, who located in remote villages of Chinese northwest zone, such as mobile vehicle of power supply, region power supply and no watch transformer substation. For instance, the DG has been used to the national defence of China. It is well known that China has more than 5000 islands, which intersperse among the 3 million km^2, and mostly the island is garrisoned by the People's Liberation Army (PLA). Thousands of PLA garrison the island in order to safeguard the coastal areas and territorial seas. But the life condition of PLA is very hardy due to the area of most of the islands are very small, and where they have not fresh water and fossil resources. Fortunately, the small islands have abundant solar and wind resources. The DG is the best way to improve the life quality of PLA and the islander. The PV DG and wind DG and solar-wind hybrid DG have been used in thousands of island army. Some other actual applications have improved the life of soldier and islander. At present, the seawater is desalted in order to provide enough drinking water, and the electric power drive thousands of martial equipment, such as radar, computer and missilery. Certainly, the standard of living is increasing by using the DG. In 1 July, 2006, the Qingzang railway is established from Xining of Qinghai province to Lhasa of Xizang by thousands of worker, and the length is 1956 km, where we have execrable environment and far-flung winter. In a word, the solar DG has great potential in future China.

4. Grid-Connected PV Generation (GPG)

At present, the grid-connect PV generation (GPG) is regarded by the developed country in the recent decades. The GPG has a biggish share in the whole yield of PV, and which will achieve a great development in future. However, the development of GPG in China is very slow, and the market share is only 0.3% in the last 30 years. The essential reason is the costly electrovalence of PV. At present, the electrovalence of PV is about 0.6 dollars per kW·h, and which is too high to support by the common people, while the electrovalence of the conventional fossil resource in China is only 0.5 RMB. Fortunately, Chinese central government had realized the problem, and some hortative policy is established, such as the generating electric power of PV must be accepted by power company, and the price is enhanced in order to ensure the advantage and enthusiasm of investors.

The desert zone of northwest China in 2007 is more than 1.3 million km². The capacity total of PV is 100 MW per km². If the fixed PV area of desert has a share of 1%, and the capacity total of PV is 13000 GW. In other words, the capacity is double compared with the accumulative total of electric power at present. With the improvement of technology and the decreasing price of PGEP, the prospect of large-scale desert PGEP is enormous in future China. At present, three PV power plants established in west desert, the capacity is more than 20 MW. The object capacity of the desert PGEP is 200 MW in 2020. For example, Yangbajing desert PGEP is established in Xizang, the capacity is 100kW.

The architecture area in China is more than 40 billion m², and housetop area is more than 4 billion m², and the area of southerly wall is more than 5 billion m². The total area can be utilized more than 49 billion m². If the fixed PV area of architecture has a share of 20%, the total capacity is 100 GW. Some actual applications of architecture PGEP have been implemented, such as solar energy demonstration city in Baoding, the international flower garden in Shenzhen and the olympic games gymnasium in Beijing.

Fortunately, the GPG have been regarded by the Chinese central government and some corporations. The biggest GPG in China lie in Dunhuang, Gansu province. The total capacity is 10 MW, the total investment is more than 73 million dollars and the area of PV is about 1 million m², and the annual accumulative total of electric power is about 16 million kW · h. The item has a short transmission distance, and the distance is about 13 km from Dunhuang city, which can provide clear energy for common people of Dunhuang. Certainly, some actual examples have been used to improve the energy structure, such as the total capacity of Chongming island item in Shanghai is 1 MW and the total capacity of Eerduosi item in Neimenggu province is 255 kW. Thus the GPG in China has a beautiful future with the increasing regard by the central government and common people.

7.3.4 Wind-Solar Hybrid System

At present, Chinese central government and local government have established more than 300 stand-alone renewable and sustainable electric stations, such as PV generation, wind generation and small water power. Especially, PV generate electricity have improved the quality of life of remote villages' people, and which increase the earning of family. However, the stand-alone PV generate electricity system has a common drawback, and the output electric power of PV is unpredictable under tremendous changes in climatic conditions. Fortunately, the hybrid system can partially overcome the problems. Some hybrid methods have been used to improve the quality of electric power, i.e., the wind-diesel, diesel solar and wind-solar hybrid systems. China has abundant wind resource in large soil, and the northwestward and the seaboard have the best wind energy. Moreover, very good compensation characters are found between solar energy and wind energy. Consequently, the hybrid system has greatly improved the quality of electric power at all time, and the higher generating capacity factors are achieved under unpredictable climatic conditions. The application of wind-solar is regarded by the common people and government, and some local government has replaced the conventional city lighting system by using wind-solar street lamp, and wind-solar hybrid DG is used to supply electric power in remote villages. However, the holding rate of wind-solar hybrid street lamp is less than the conventional street lamp at present. The primary reason is the costly price of hybrid system. For example, the price of wind-solar hybrid street lamp is more than 4000 dollars, and the price of conventional street lamp is about 2200 dollars at the same time. Certainly, the price will descend with the improvement of technology, and the wind-solar hybrid system still has great potential in future.

7.4 Nuclear Power and Other

7.4.1 How Nuclear Power Works

A nuclear power plant produces electricity in almost exactly the same way that a conventional (fossil fuel) power plant does. A conventional power plant burns fuel to create heat. The fuel is generally coal, but oil is also sometimes used. The heat is used to raise the temperature of water, thus causing it to boil. The high temperature and intense pressure steam that results from the boiling of the water turns a turbine, which then generates electricity. A nuclear power plant works the same way, except that the heat used to boil the water is produced by a nuclear fission reaction using uranium-235 (U-235) as fuel, not the combustion of fossil fuels. A nuclear power plant uses much less fuel than a comparable fossil fuel plant. A rough estimate is that it takes 17000 kg of coal to produce the same amount of electricity as 1 kg of nuclear uranium fuel.

The nuclear power plant stands on the border between humanity's greatest hopes and its deepest fears for the future. On one hand, atomic energy offers a clean energy alternative that frees us from the shackles of fossil fuel dependence. On the other hand, it summons images of disaster: quake-ruptured Japanese power plants belching radioactive steam, the dead zone surrounding Chernobyl's concrete sarcophagus.

But what happens inside a nuclear power plant to bring such marvel and misery into being? Imagine following a volt of electricity back through the wall socket, all the way through miles of power lines to the nuclear reactor that generated it. You'd encounter the generator that produces the spark and the turbine that turns it. Next, you'd find the jet of steam that turns the turbine and finally the radioactive uranium bundle that heats water into steam. Welcome to the nuclear reactor core.

The water in the reactor also serves as a coolant for the radioactive material, preventing it from overheating and melting down. In March 2011, viewers around the world became well acquainted with this reality as Japanese citizens fled by the tens of thousands from the area surrounding the Fukushima-Daiichi nuclear facility after the most powerful earthquake on record and the ensuing tsunami inflicted serious damage on the plant and several of its reactor units. Among other events, water drained from the reactor core, which in turn made it impossible to control core temperatures. This resulted in overheating and a partial nuclear meltdown.

As of March 1, 2011, there were 443 operating nuclear power reactors spread across the planet in 47 different countries. In 2009 alone, atomic energy accounted for 14% of the world's electrical production. Break that down to the individual country and the percentage skyrockets as high as 76.2% for Lithuania and 75.2% for France. In the United States, 104 nuclear power plants supply 20% of the electricity overall, with some states benefiting more than others.

A typical PWR nuclear power unit with 983.8 MW power output in China is selected for study. Fig. 7.12 shows the schematic diagram of steam cycle of the power unit. It is a conventional reheat-regenerative Rankine cycle. The thermodynamic states of different nodes in the cycle are indicated by open circles with numbers, while the thermal components are identified using the corresponding designations. The key components of the steam cycle include a pressurized water reactor (PWR), three steam generators (only one is shown in the diagram for clarity), a high pressure and three low pressure steam turbines, a generator connecting to the turbines by a common shaft. The heat generated by the nuclear fuel in the PWR is transferred to the steam generator system (SGS), where the feedwater is heated and then boiling happens. The steam generated by the SGS is almost at saturated state of 6.71 MPa and 283℃, and has mass flow rate of 1613.4 kg/s. It flows into the turbines and then expands to generate power in the turbine train.

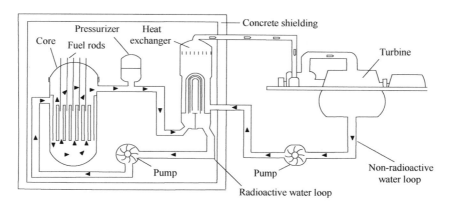

Fig. 7.12 PWR power plant schematic

7.4.2 Nuclear Fission: The Heart of the Reactor

Despite all the cosmic energy that the word "nuclear" invokes, power plants that depend on atomic energy don't operate that differently from a typical coal-burning power plant. Both heat water into pressurized steam, which drives a turbine generator. The key difference between the two plants is the method of heating the water.

While older plants burn fossil fuels, nuclear plants depend on the heat that occurs during nuclear fission, when one atom splits into two and releases energy. Nuclear fission happens naturally every day. Uranium, for example, constantly undergoes spontaneous fission at a very slow rate. This is why the element emits radiation, and why it's a natural choice for the induced fission that nuclear power plants require.

Uranium is a common element on Earth and has existed since the planet formed. While there are several varieties of uranium, uranium-235 (U-235) is the one most important to the production of both nuclear power and nuclear bombs.

U-235 decays naturally by alpha radiation: It throws off an alpha particle, or two neutrons and two protons bound together. It's also one of the few elements that can undergo induced fission. Fire a free neutron into a U-235 nucleus and the nucleus will absorb the neutron, become unstable and split immediately.

When a uranium-235 nucleus with a neutron approaching from the top, as soon as the nucleus captures the neutron, it splits into two lighter atoms and throws off two or three new neutrons (the number of ejected neutrons depends on how the U-235 atom splits). The process of capturing the neutron and splitting happens very quickly.

The decay of a single U-235 atom releases approximately 200 MeV (million electron volts). That may not seem like much, but there are lots of uranium atoms in a pound (0.45 kilograms) of uranium. So many, in fact, that a pound of highly enriched uranium as used to power a nuclear submarine is equal to about a million gallons of gasoline.

The splitting of an atom releases an incredible amount of heat and gamma radiation, or radiation made of high-energy photons. The two atoms that result from the fission later release beta radiation (superfast electrons) and gamma radiation of their own, too.

But for all of this to work, scientists have to first enrich a sample of uranium so that it contains 2% to 3% more U-235. Three-percent enrichment is sufficient for nuclear power plants, but weapons-grade uranium is composed of at least 90% U-235.

7.4.3 Pros and Cons of Nuclear Power

What's nuclear power's biggest advantage? It doesn't depend on fossil fuels and isn't affected by fluctuating oil and gas prices. Coal and natural gas power plants emit carbon dioxide into the atmosphere, which contributes to climate change. With nuclear power plants, CO_2 emissions are minimal.

According to the Nuclear Energy Institute, the power produced by the world's nuclear plants would normally produce 2 billion metric tons of CO_2 per year if they depended on fossil fuels. In fact, a properly functioning nuclear power plant actually releases less radioactivity into the atmosphere than a coal-fired power plant. In addition, all this comes with a far lighter fuel requirement. Nuclear fission produces roughly a million times more energy per unit weight than fossil fuel alternatives.

And then there are the negatives. Historically, mining and purifying uranium hasn't been a very clean process. Even transporting nuclear fuel to and from plants poses a contamination risk. And once the fuel is spent, you can't just throw it in the city dump. It's still radioactive and potentially deadly.

On average, a nuclear power plant annually generates 20 metric tons of used nuclear fuel, classified as high-level radioactive waste. When you take into account every nuclear plant on Earth, the combined total climbs to roughly 2000 metric tons a year. All of this waste emits radiation and heat, meaning that it will eventually corrode any container that holds it. It can also prove lethal to nearby life forms. As if this weren't bad enough, nuclear power plants produce a great deal of low-level radioactive waste in the form of radiated parts and equipment.

Over time, spent nuclear fuel decays to safe radioactive levels, but this process takes tens of thousands of years. Even low-level radioactive waste requires centuries to reach acceptable levels. Currently, the nuclear industry lets waste cool for years before mixing it with glass and storing it in massive cooled, concrete structures. This waste has to be maintained, monitored and guarded to prevent the materials from falling into the wrong hands. All of these services and added materials cost money—on top of the high costs required to build a plant.

VOCABULARY

1. polymer   n. 聚合物
2. straw   n. 稻草；吸管；一文不值的东西
3. biodiesel   n. 生物柴油
4. synthetic   adj. 综合的；合成的，人造的
5. diesel   n. 柴油机；柴油   adj. 内燃机传动的；供内燃机用的
6. combustion   n. 燃烧，氧化
7. pulverized   adj. 粉状的；呈粉末状的
8. fission   n. 裂变；分裂
9. prognosis   n. 预测
10. geothermal   adj. 地热的
11. tutorial   n. 专题报告，专题论文
12. nacelle   n. 短舱，引擎舱
13. gearbox   n. 齿轮箱；变速箱
14. kinetic   adj. 动力的；运动的
15. aerodynamic   adj. 空气动力学的，航空动力学的
16. strategy   n. 战略，策略
17. momentum   n. 动量；动力
18. teeter   vt. 使⋯摇摆；使⋯上下晃动
19. hinge   n. 转轴；铰链
20. discrepancy   n. 相差；矛盾；不符
21. deviation   n. 差异，偏差
22. yaw   n. （火箭、飞机、宇宙飞船等）偏航   vt. 使⋯偏航
23. nocturnal   adj. 夜的；夜间发生的
24. diurnal   n. 日记账；日报，日刊
25. portfolio   n. 业务量；业务责任；公文包
26. turbulent   adj. 骚乱的，混乱的；狂暴的
27. anemometer   n. 风力计，风速计
28. sonic   adj. 音波的；声音的
29. hinder   vt. 阻碍；打扰
30. align   vt. 使成一行；匹配；使结盟
31. gyroscopic   adj. 回转仪的，陀螺的
32. prevail   vi. 盛行，流行；战胜
33. lull   vt. 使平静；使安静
34. photovoltaic   adj. ［电子］光电池的，光电的
35. criterion   n. 标准；准则；规范
36. gallium   n. ［化学］镓
37. arsenide   n. 砷化物
38. indium   n. ［化学］铟
39. phosphide   n. 磷化物
40. cadmium   n. ［化学］镉
41. telluride   n. 碲化物
42. diselenide   n. 联硒化物
43. amorphous   adj. 非晶形的；无定形的；无组织的

44. alloy　　*n.* 合金　　*vt.* 使成合金
45. susceptible　　*adj.* 易受影响的
46. epoxy　　*n.* 环氧基树脂
47. fiberglass　　*n.* 玻璃纤维；玻璃丝
48. siphon　　*n.* 虹吸管；虹吸
49. livestock　　*n.* 家畜；牲畜
50. herd　　*n.* 兽群，畜群；放牧人
51. electrovalence　　*n.* 电价
52. hortative　　*adj.* 奖励的；劝告的
53. uranium　　*n.*［化学］铀
54. coolant　　*n.* 冷却剂
55. spontaneous　　*adj.* 自发的；自然的；无意识的
56. photon　　*n.* 光子；辐射量子

NOTES

1. In promotion of the renewable energy policy, the large scale expansion of constructions of the wind and solar power is developing in China, whose random high-power electric energy fluctuation causes huge shock to the power grid.

在新能源政策的推动下，中国的风能和太阳能发电站得以大规模扩建，其随机的大功率电能波动给电网带来巨大冲击。

2. We then explain the layout of a wind turbine control system by taking a "walk" around the wind turbine control loop, with discussions on wind inflow characteristics, available sensors and actuators, and turbine modeling for use in control.

然后我们沿着其控制回路来解释风力涡轮机控制系统的布局，并讨论风的流入特性、传感器和执行器以及控制用的涡轮机模型。

3. Variable-speed operation can also reduce turbine loads, since sudden increases in wind energy due to gusts can be absorbed by an increase in rotor speed rather than by component bending.

由于因阵风引起的风能突然增加可以由风机转子的加速吸收掉而不会令风机部件弯曲，因此，变速运行可以减少风力涡轮机的负荷。

4. Both turbines generate the same power at the fixed speed turbine's 10 m/s design point, but the variable speed turbine generates more power at all wind speeds in Region 2, which with a maximum difference of about 150 kW at 6 m/s.

虽然定速10m/s的设计点时两个风力涡轮机产生相同的功率，但是变速涡轮风力发电机在区域2各种风速下都能产生更多功率，当风速为6m/s时二者功率最大差值达150 kW。

5. Calibration drift is a typical failure mode, especially for strain gages and accelerometers, and thus controllers that depend on sensors prone to drift must be robust to calibration errors such as unknown biases.

特别是对于压力应变计和加速度计来说，校准漂移是一种典型的失效模式。因此依赖于易漂移传感器的控制器在如未知偏差的校正误差方面必须具有鲁棒性。

6. Wind turbines are often located with other turbines in wind farms to reduce costs by taking advantage of economies of scale.

在风电场中，风力涡轮机通常和其他涡轮机一起安装，以利用规模效应降低成本。

7. Shorter cross wind spacing distances can reduce land costs and are beneficial at locations

where the frequency distribution of the wind direction is skewed heavily toward the prevailing direction.

较短的横风间距可以降低土地成本，且在风向分布经常性严重偏离主风向的地方尤其有益。

8. The spatial variation of turbines on a wind farm often results in power smoothing, where the standard deviation of the power produced by multiple turbines is less than the standard deviation of the power produced by each individual turbine.

风电场中，涡轮机位置的空间变化使风电场的功率更加平稳，因为多个风力涡轮机产生的功率标准偏差小于单个风力涡轮机所产生的功率标准偏差。

9. This large area must be produced and deployed at as low a cost as possible and the cells should ideally be made by a low-cost mass-production process.

这种大区域必须在尽可能低的成本下生产和部署，理想的电池组生产方式是低成本的、大规模的生产。

10. The energy consumption of an active cooling system would not be compensated by the gain in increased energy generation, at least for small systems.

至少对于小型系统来说，增加的发电量可能不够补偿主动冷却系统的能量消耗。

11. The annual total amount of whole city road lighting system is more than one billion kW·h, and billions of dollars is expended in order to pay the expense of electric power, which had became the enormous burden of the Chinese government.

城市道路照明系统每年耗电总量超过十亿千瓦时，数十亿美元用以支付电力开支，这已经成为中国政府的巨大负担。

12. However, the stand-alone PV generate electricity system has a common drawback, and the output electric power of PV is unpredictable under tremendous changes in climatic conditions.

然而，独立光伏发电系统有一个普遍的缺点，当气候条件变化剧烈时光伏输出的电能是不可预测的。

Chapter 8
Smart Power Grid

8.1 Introduction

The former U. S. president Barack Obama calls for smart grid in his energy speeches. Al Gore says it's critical to repower America. The U. S. Department of Energy says it will transform the nation's electrical system the way the internet did to the way we live, work, play and learn. But what exactly is the smart grid? Think of it as a modernized, more efficient way to distribute electricity to save energy and money. Right now, the electricity we get tends to come in one direction: Power plants generate power and it is distributed to the nation's homes and businesses through the electric grid.

But those homes and business can't communicate very well with the system, letting them know when they are over-burdened or when power outages occur. Utility officials often have to rely on people to call them to let them know when the lights are out and can't "see" into the system very well to spot potential problems with demand and supply. When outages occur, repairmen often have to spend hours or more than a week, if the recent storm outages in Fitchburg are any indication, tying homes and businesses back into the grid.

A smart grid will use digital technology to allow two-way communication between electricity generators and customers. It will, if promises hold true, allow appliances in homes to use electricity when it is abundant and inexpensive. It will allow electricity managers to peer into their systems to identify problems and avoid them. It will provide rapid information about blackouts and power quality.

It is also designed to link the grid to large scale solar and wind projects that are built far away from the cities and suburbs where people need electricity. Through these transmission lines and better management of the intermittent power from wind, for example, it will allow the nation to incorporate large scale renewable power into the power system.

It will also allow a community to use local power sources to keep the lights on even when there is no power coming from a utility. Called islanding, it will allow a home to grab power from "distributed resources"—local rooftop solar, small hydro and wind projects for example—to keep the lights on until utility workers can patch the community back into the grid. At the same time, the entire nation of electricity distribute will become interconnected.

"The former U. S. president Barack Obama wants 10% of electricity to be renewable. And, our current grid infrastructure is a barrier to meeting that goal. A smart grid is crucial for sending electronic messages through the system for power generation triggers and demand reduction calls," says Vincent DeVito, a former U. S. Assistant Secretary of Energy for Policy and International Affairs who is now a partner at Bowditch & Dewey LLP in Boston.

Now this is a really ambitious proposal. The Department of Energy says it's at least 10 years away. Others say more. It will cost billions of dollars. Yet DeVito and others say it's critical to do and will save money in the long run. The Department of Energy says if the grid were just 5% more efficient, the energy savings would equate to eliminating the fuel and greenhouse gas emissions from 53 million cars.

This chapter will introduce the concepts and architecture of the smart substation and the smart grid.

8.2 Smart Substation

8.2.1 Introduction

Many transmission substations in service rated 110 kV and above in the USA, are older than 40 years. They had often been built in several stages. It is usual to find in the same substation equipment belonging to different technology vintages and different manufacturers. The maintenance and operation cost is high due to the legacy devices. Besides, the legacy power apparatus may have potential safety and environment issues. When building a new substation, which does not happen very often in the USA, one has an opportunity to use prior experiences when deciding on the requirements of the new design.

The conventional air-insulated substation (AIS) design uses a large number of disconnectors in order to allow for maintenance and repair with a minimum of interruption. The occupied area of AIS is typically large and the maintenance demand of the open-air apparatus is relatively high, particularly in case of severe environmental conditions. Besides, switchgear, its subsystems and components are exposed to aging and wearing during the years of exploitation that leads to the increase in fault events over the years of service. The attempt in the new substation designs is to make them more compact and somewhat protected from the environmental impacts.

The sensing and signal processing in existing substation designs is based on a number of individual sensors being placed in the switchyard and hardwired directly to the control house. The individual monitoring, control and protection devices that are using those signals for their decision-making are located in the control house. This concept is not facilitating integration of data and signal processing across the substation.

The IEC 61850 substation automation standard provides higher degree of integration, greater flexibility, and reduces construction and commissioning time. The levels of functional integration and flexibility of communications bring significant advantages in cost reduction. This integration affects not only the design of the substation but almost every component and/or system such as protection, monitoring and control by allowing replacement of the hardwired interfaces with communication links.

The new primary equipment design needs to be compact, environmental friendly and allow low cost of operation and maintenance. The new secondary side design is based on IEC 61850 standards and needs to utilize synchronized sampling technology and multifunctional IEDs. The proliferation of vulnerability of protection, control and automation systems using switched Ethernet communications between devices and between substations requires that the cyber security issue also be emphasized.

This section first covers the primary equipment design, then the hardware and software implementation of the secondary equipment design, and finally the benefits of the new design.

8.2.2 Primary Equipment Design

1. Gas Insulated Substation (GIS) Design

The metal-enclosed gas-insulated switchgear inherently follows the criteria for new substation design and offers a higher reliability and flexibility than other solutions. Due to the gas enclosed design, GIS is the most suitable solution for indoor and underground substations. In outdoor and hybrid substations, the occupied area is tremendously reduced by using GIS technology.

GIS configurations can be applied to any type of bus bar arrangements: single busbar, double busbar, single busbar with transfer bus, double busbar with double circuit breaker, one and a half circuit breaker scheme and ring busbar. One line diagram of double busbar is depicted in Fig. 8.1. Fig. 8.2 shows the layout of a GIS substation based on one line diagram in Fig. 8.1.

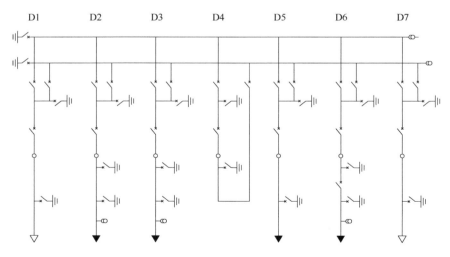

Fig. 8.1 One line diagram of double busbar

In Fig. 8.2, it may be observed how the compact design of GIS reduces substation area tremendously (at least 70%) compared to the same AIS configuration. This fact allows GIS to become the choice of preference for indoor and underground substation. For a better appearance, an underground GIS substation can be design with an aesthetic view that hides its presence.

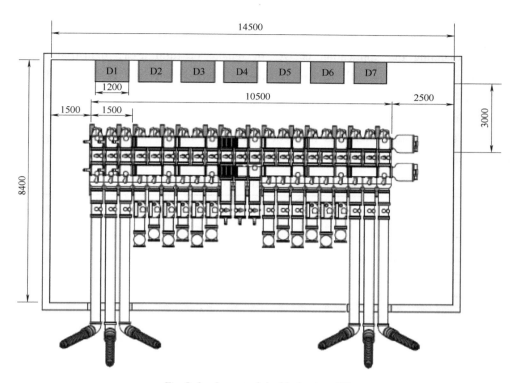

Fig. 8.2 Layout of double busbar GIS

The existence of a substation could be designed so that it cannot even be recognized, such as given in example in Fig. 8.3.

Fig. 8.3 Overview of an 500 kV underground GIS substation

GIS performs the same function as AIS. The compact and metal-enclosed design of GIS has prominent advantages and better performance than AIS. However, the high initial investment is a key obstacle in expanding the application of GIS. In remote or rural area, industrial areas or in developing countries, AIS is still the best choice. In places where the cost of land or cost of earthworks is high or where the sceneries cannot be disturbed by AIS, the solution is to use underground or indoor GIS.

Different type of substations has different advantages which come from its components and design. The characteristics of GIS and AIS are given in the Tab. 8.1.

Tab. 8.1 Design characteristic of GIS and AIS

Characteristics	Air-insulated switchgear		Gas-insulated switchgear	
	AIS	Dead-tank	HIS	GIS
Type of installation	Outdoor	Outdoor	Outdoor	Indoor
Metal-encapsulated circuit breaker	-	×	×	×
Metal-encapsulated disconnector	-	-	×	×
Metal-encapsulated earthing switch	-	-	×	×
Busbar	Air-insulated	Air-insulated	Air-insulated	Gas-insulated
SF_6 insulated current transformer	-	-	×	×
SF_6 insulated voltage transformer	-	-	×	×
Direct cable of SF_6/oil termination	-	-	×	×

From the Tab. 8.1, the difference in design characteristics may be observed. As may be noted, the GIS designs are supposed to be applied where one or more of the following features are desirable: limited space, extreme environmental conditions, required low environmental impact and less maintenance. The low failure rate of GIS is also a prominent advantage. But the outage time (56 h), double outage time of AIS (25 h), is one of the disadvantages.

Regarding economics, initial capital investment is not enough to evaluate the overall substation project. Life cycle cost (LCC) should be considered, including primary hardware cost, maintenance cost, operation cost, outage cost and disposal costs. The LCC comparison of AIS and GIS is as follows:

1) Primary hardware: For primary equipment, GIS is more expensive than AIS. However, the price of auxiliary equipment such as support, conductors, land, installation, control, protection and monitoring can lead to a cost difference between the two systems being small.

2) Maintenance: The failure rate of circuit breaker and disconnecting switch in GIS is one-fourth of that of AIS and one tenth in case of busbar, thus the maintenance cost of GIS is less than that of AIS over the lifetime.

3) Operation cost: The maintenance cost of GIS and AIS shall be equivalent. The cost for training in GIS is higher than in AIS.

4) Outage cost: Since the failure rate of GIS is lower, the outage cost of AIS shall be greater.

5) Disposal cost: The cost of decommissioning and disposal after use should be capitalized. The value of future expense must be taken into account.

The general conclusion about the LCC advantages of AIS versus GIS cannot be easily reached; hence it can only be determined in specific project. An example below illustrates the LCC comparison is shown in Tab. 8.2. In this example, GIS and AIS use H-configuration with three circuit breakers. Fig. 8.4 shows the design of AIS and GIS solutions.

Tab. 8.2 LCC evaluation of AIS and GIS (based on calculation conducted by a utility)

Life cycle cost	AIS (%)	GIS (%)
Planning and engineering	100	80
Real estate	100	40
Primary equipment	100	120
Secondary equipment	100	100
Earthwork, civil work, structures	100	60
Electrical assembly and erection	100	70
Maintenance	100	50
Outage	100	50
LCC after 10 years	100	max 70

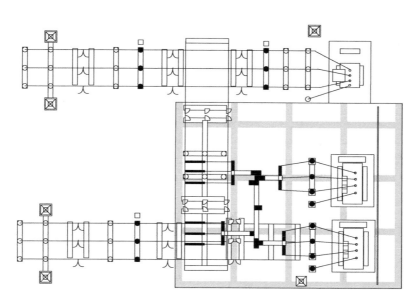

Fig. 8.4 GIS and AIS solutions in H-configuration

The latest GIS technology has less environmental impact than previous technology. The SF_6 leakage rate is less than 1% (in experiment <0.5%). Due to the design characteristics, GIS has a better impact on environment than AIS. Environment impact comparison of AIS and GIS is shown in Tab. 8.3.

Tab. 8.3 Environment impact comparison of AIS and GIS

Items	AIS (%)	GIS (%)
Primary energy consumption	100	73
Area requirement	100	14
Acidification potential	100	81
Greenhouse potential	100	79
Nitrification potential	100	71

As observed, GIS offers many prominent advantages over AIS. It meets all requirements for new substation design, except high initial investment and potential environmental risk. This disadvantage and the ever increasing environmental awareness become the drivers for a new generation of GIS that complies with future green field substation criteria.

2. Disconnecting Circuit Breaker (DCB) design

In order to evaluate the benefits of DCB, a typical 230/69 kV is selected to compare DCB and conventional combination of circuit breaker and disconnecting switch. Comparison of DCB substation and original substation areas is shown in Fig. 8.5.

The specification of the equipment in the substation is as follows:

1) Transformer: 240 MV·A, 230/69kV;
2) Circuit breaker: 230 kV, 3000A and 69 kV, 3000A;
3) Disconnecting switch: 230 kV, 3000A and 69 kV, 3000A.

The cost between DCB and conventional combination of circuit breaker and disconnecting switch is compared in Tab. 8.4. The main features between two types of solutions in the example of specific layouts of the two substation types may be observed.

Tab. 8.4 Comparison between DCB and conventional circuit breaker

Items	DCB		Conventional	
	230 kV	69 kV	230 kV	69 kV
# of circuit breaker	15 × 3	26 × 3	15 × 3	26 × 3
# of disconnector	0	0	27 × 3	51 × 3
# of foundation	74		127	
Dimension	55 m × 48 m	24 m × 84 m	96.3 m × 80.5 m	26.8 m × 127.7 m

Device	Conventional			DCB		
	Number	Cost/ $	Total/ $	Number	Cost/ $	Total/ $
Circuit breaker	41	111400	4567400	41	185000	7380000
Disconnecting switch (3-phase)	78	11000	858000	0	11000	0
CT, PT	165	20000	3300000	165	20000	3300000
Foundation	127	10000	1270000	74	10000	740000
Total			9995400			11420000

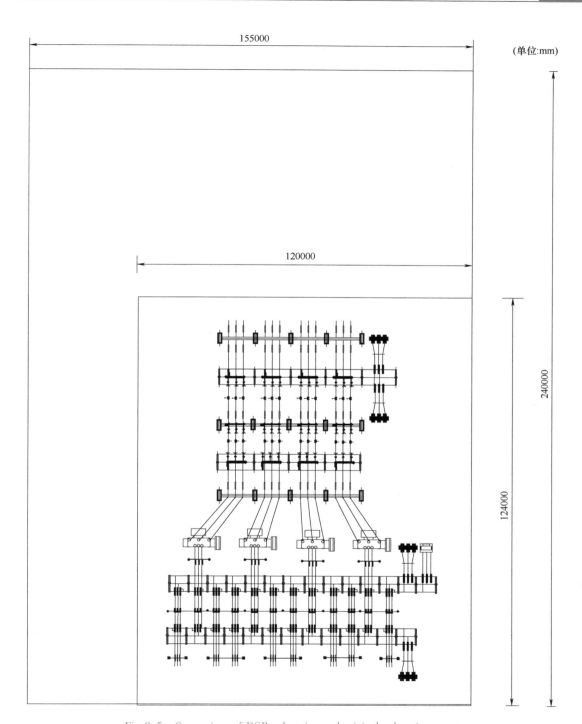

Fig. 8.5　Comparison of DCB substation and original substation areas

In this evaluation, only the 230 kV and 69 kV switchgear is taken into account, because other primary parts are similar for both solutions. By omitting disconnecting switch when using DCB in switchgear, the substation can be much smaller and more cost effective. 20% to 50% of space requirement is saved.

The cost of conventional circuit breaker, disconnecting switch, CT, PT, foundation are provided by US Grid. The cost of disconnecting circuit breaker and auxiliary part are provided by ABB in reference. The costs of design and planning, civil work, busbar and connection, and failure and maintenance have to be considered in the overall cost.

Overall cost difference between switchgear using DCB and conventional switchgear is presented in Tab. 8.5.

Tab. 8.5 Cost comparison

Items	DCB/$	Conventional/$
Primary apparatus	11420000	9995400
Failure and maintenance	526300	1052600
Busbar and connections	2368300	2368300
Civil work and sitework	11420000	16841600
Design and planning	8420800	10526000
Total	34155400	40783900

From Tab. 8.5, the cost of switchgear using DCB is 85% of the conventional. This is just estimated cost so the saving is in 5% to 15% range. The cost saving from land deduction (20% to 50%) would also add more to the overall saving percent, especially at places where land cost is high.

8.2.3 Secondary Equipment Design

The solutions proposed in this section utilizes advanced concept for data and information processing. The copper wiring between IEDs and power apparatus in conventional substations is replaced with optical fibers. All the multifunctional IEDs in the control house fully support IEC 61850 protocol. The recorded data is synchronized and time stamped utilizing receivers for global position system (GPS) or computer network synchronization (IEEE PC 37.238). The time reference signal is distributed to downstream devices or IEDs by Inter Range Instrumentation Group (IRIG) standard signal or IEEE 1588 V2 signal. Hardened Ethernet switches are used to process the message priority to realize the generic object oriented substation event (GOOSE) messaging scheme between relays and provide security at the local area network level.

The software applications include automated analysis of data from IEDs such as circuit breaker monitor (CBM), digital fault recorder (DFR), digital protective relay (DPR) and other substation IEDs. The new system enables automated collection of field data, extraction of information and sharing of information among different utility groups allowing them to have better view of the system.

1. Hardware Implementation

The overall system architecture is show in Fig. 8.6.

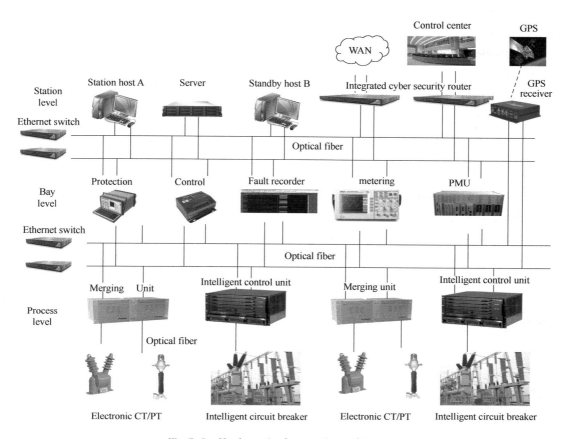

Fig. 8.6 Hardware implementation architecture

(1) Process Level Solution

Optical current and voltage sensors are used instead of conventional transformers. The voltage and current signals are connected at the primary side, converted to the optical signals by merging unit and transferred to the protection and control devices in the control house via optical fibers. This can lower the requirement of transformer insulation and reduce the interference present in the analog signal transmission. Intelligent control units are used as an intermediate link for circuit breaker interfacing. The intelligent control unit converts analog signals from primary devices (such as circuit breaker and switches) into digital signals and sends it to the protection and control devices via process bus. At the same time, the tripping and reclosing commands issued by protection and control devices are converted into analog signals and sent to the switchyard to control the primary equipment. Large amount of copper wiring between IEDs and primary devices in conventional substations is replaced by optical fibers.

(2) Bay Level Solution

All the IEDs in the control house fully support IEC 61850 protocol. Synchronous phasor measurements are realized by phasor measurement units (PMUs). PMUs are used for wide area power system monitoring and control. The interoperation between IEDs is realized by using GOOSE massage network. The Ethernet switch is used to process the message priority to realize the GOOSE data exchange scheme between relays.

(3) Station Level Solution

Manufacturing messaging specification (MMS) network, which is the communication link between SCADA, control center and IEDs at bay level is used at the station level. Redundant networks are used for high reliability.

(4) Cyber Security Issues

Since the Ethernet switches and routers are used in the network, the major concern is the cyber security. Hardened routers are used to specifically provide an electronic security perimeter for the protection of critical cyber assets. The hardened switches are used to provide security at the local area network level.

Besides the hardened routers and switches, secured access control device is used in the new design to further protect the security of the substation. The main function of such control device is shown in Fig. 8.7.

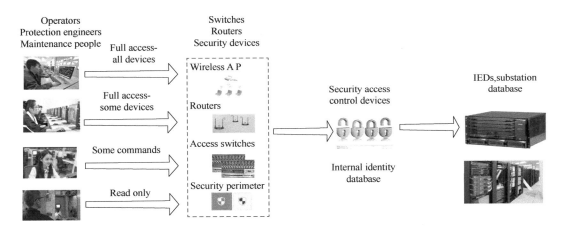

Fig. 8.7 Function of security access control device

There are four levels of access to the substation database. Operators are allowed a selected access (includes read and write) to all the devices (relays, meters and controllers). Protection engineers have the full access to some devices (protective relays). Maintenance staff can only have the partial access to the devices. Customers or utilities can only read substation's information. According to the different access levels, the different passwords or secured access methods are specified.

(5) Time Synchronization Methods

Modern protection, monitoring and control systems rely on the availability of high accuracy time signal. Time synchronization eliminates or reduces the effort involved in correlating event information from distributed intelligent disturbance recording devices. Accurate time signal is required for disturbance analysis in order to correlate individual device event reports and is essential for synchronized system control and synchrophasor utilization. Moreover, time synchronization is also required when integrating data from different IEDs in different locations. GPS is the most common source to provide high accuracy time signal in current substations. Hence, the GPS becomes the single point of failure caused for example by solar activity, intentional or unintentional jamming, or U. S. Department of Defense (DOD) modifying GPS accuracy or turning off the satellite system.

IEC 61850 recommends the Network Time Protocol (NTP) as the primary synchronization method, but NTP time accuracy is insufficient for Sampled Values (SV) applications (< 1 μs). IEEE 1588 Standard for a Precision Clock Synchronization Protocol for Networked Measurement and Control Systems may be adopted to distribute the 1 μs timing accuracy. IEEE 1588 uses the LAN cables to distributed high accuracy time, eliminating the need for the additional IRIG cabling compared to separate IRIG time reference distribution. Clocks that support the IEEE 1588 standard have more options for alternate time sources which will provide time source redundancy.

Rugged communications devices (RSG 2288 and RS 416) suggested in the new design are capable of receiving IEEE 1588 V2 through their Ethernet ports and distributing IEEE 1588 V2 or generating synchronized IRIG-B signal for legacy devices. Those devices used in the design make implementation of the IEC 61850-9-2 "Process Bus" more economical, more practical and easier to deploy by providing reliable and precise time synchronization over the substation Ethernet network. Besides, the new design facilitates the migration path from legacy solutions and paves the way towards IEC 61850 Edition 2.

2. Software Implementation

In order to integrate data from multiple substation IEDs and extract information for different utility group use, a software package, which can combine bulk of data recorded by individual substation IEDs such as DPRs, DFRs, PMUs and CBMs and streamline it to provide a more relevant and versatile source of information to serve control center applications is used.

Fig. 8.8 shows different levels of software-based analysis. The system wide analysis includes fault analysis and fault location (FAFL). The substation level analysis includes Digital Fault Recorder Analysis (DFRA), Circuit Breaker Monitor Analysis (CBMA), Power Quality Monitor Analysis (PQMA) and Digital Protection Relay Analysis (DPRA). The monitoring/ tracking level includes: Verification of Substation Database (VSDB), Two-stage State Estimator (TSSE), Substation Switching Sequence Verification (SSSV) and the Identification of Substation Database (ISDB). In this section, the substation level analysis is of main concerned. The substation level analysis provides information to serve control center applications such as fault location, topology processor for state estimator and alarm processor. This analysis together with power system component models can provide system with wide level disturbance monitoring and analysis solution.

The substation data is divided into two categories: non-operational and operational data. The software discussed here is adding the nonoperational data from relays, recorders and PMUs to SCADA data, because redundancy of data is very important in data analyzing and decision-making process. The new data processing uses standardized data and communication formats.

Fig. 8.9 shows a simplified diagram of the substation level integration of data. Data is collected from IEDs in Common Format for Transient Data Exchange (COMTRADE), and if not, the native

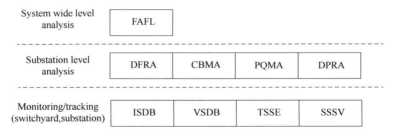

Fig. 8.8　Software implementation architecture

format is changed to COMTRADE first, and then the data is processed at the substation level and populated into the database together with the automated analysis reports and recording system configuration information.

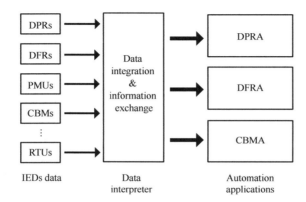

Fig. 8.9　Substation level data integration

Three types of the automated analysis applications: CBMA, DPRA, and DFRA will be discussed briefly here. IEDs are synchronized to the GPS reference clock and time stamped data makes integration of data from different IEDs much easier. IEDs from various vendors may have different data files formats, so it is necessary to standardize file format before data integration.

(1) CBMA

CBMA is an application based on analysis of records of waveforms taken from the circuit breaker using a circuit breaker monitor (CBM), explains event, and suggests repair actions. CBMA uses advanced signal processing and expert system techniques to enhance speed and provide timely results that are consistent. It enables protection engineers, maintenance crews and operators to quickly and consistently evaluate circuit breaker performance, identify performance deficiencies and trace possible reasons for malfunctioning. The event report includes several sections: The first section provides the date and time when event occurred, as well as general information regarding device; the second section (signal processing log) provides information regarding analysis operation and if there is no problem in data processing this area is empty; the third section (expert system log) provides information about signals affected by tripping operation and points out abnormities; the last section (maintenance and repair operation log) suggests possible actions to be taken in repairing the device. The main function of CBMA is shown in Fig. 8.10.

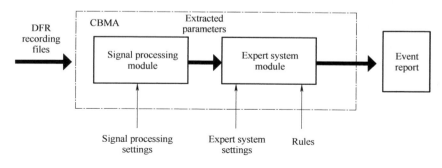

Fig. 8.10 CBMA function diagram

(2) DPRA

DPRA is expert system based analysis software which automates validation and diagnosis of digital protective relay (DPR) operation. It takes various relay reports and files as inputs and using embedded expert system generates a report on the results of analysis. Validation and diagnosis of relay operation is based on comparison of expected and actual relay behavior in terms of status and timing of logic operands. The analysis report summarizes general fault information such as fault inception/clearance time, fault type and location and lists logic operands and notifies their status and operating sequence. If some operand failed to operate, the verifying action will be suggested. The main function of DPRA is shown in Fig. 8.11.

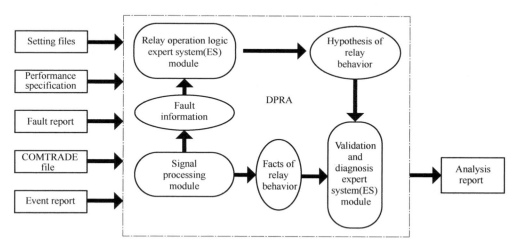

Fig. 8.11 DPRA function diagram

(3) DFRA

DFRA provides automated data analysis and integration of digital fault recorder (DFR) event records. It provides conversion from different DFR file formats to COMTRADE. Besides, DFRA performs signal processing to identify preand post-fault analog values, statuses of the digital channels (corresponding to relay trip, breaker auxiliary, communication signals), fault type, and faulted phases. It also checks and evaluates system protection, fault location, etc. The report consists of several sections: the first section (expert system log) displays event summary and time (in cycles) required for device to act in detecting faulted line and clearing fault; the second section (event origin) displays affected circuit and substation; the third section (event summary) provides general information about fault clearing and device operation; the fourth section (analog signal values) display pre-, during and post-fault value of voltage and current; the last section (digital signal status) displays time of device trip operations. The main function of DFRA is shown in Fig. 8.12.

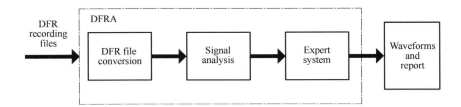

Fig. 8.12 DFRA function diagram

8.2.4 Benefits of the New Substation Design

The benefits of the primary equipment design are as follows:

1) The GIS has the following benefits: higher operational safety, higher availability and reliability, lower maintenance requirement, lower environmental impact, lower operating cost, ability to install where space is limited.

2) DCB based primary equipment design has the following benefits: saving equipment cost, reducing the footprint and related construction cost, increasing availability.

The benefits of the secondary equipment design are as follows:

1) Cost: Optical fiber replaces costly copper wires, which saves investment money, and reduces the construction labor and maintenance cost.

2) Reliability: Redundant design improves the reliability of the whole system, and some level of self-healing can be realized by automatic transfer scheme between different IEDs which could be done by GOOSE messaging.

3) Data sharing: Automatic data analysis software used in the control house allows data sharing among different utility groups improving their understanding of the events.

4) Cyber security: Hardened switches and routers provide reliable security.

8.2.5 Future Substation Design

1. Introduction

The main features of a future substation are high reliability, economical benefit, simplicity, intelligence, modularization and low environmental impact. Driven by these requirements, as well as significant technology developments, a new concept for the substation of the future will emerge.

Designing the substation of the future will require an understanding of interaction between the primary and secondary equipment in the substation, the transformation of primary system parameters to secondary quantities used by multifunctional IEDs, and the availability of new types of sensors that eliminate many of the issues related to conventional instrument transformers.

The substation of the future will be based on modular approach to the design of the substation primary system, the multifunctional IEDs provide protection, control, measurements, recording and other functions, as well as their integration in substation automation systems with advanced functionality. The electricity markets are being restructured to become more competitive and to facilitate bulk power transfers across wider geographical regions. A critical implication of this restructuring will be to make electricity markets even more intensely data driven, creating a need for better ways of monitoring market activity in real time and sharing information among market participants. Future "smart" substations will be capable of providing such information. In the future power system, electrical events affect not only the operation of the power system, but also operation of the electricity market. It can be conjectured that the importance of an electric event should consider the economic importance of the event, and the economic impact should be taken in consideration when electrical alarms occur.

Therefore, it is proposed that alarm issuance and alarm processing should include economic information in addition to the traditional alarms. In this section, an intelligent economic alarm processor (IEAP) structure that combines alarm processing techniques at both the substation and control center level will be presented.

2. Primary Equipment Design

(1) High Temperature Superconductors (HTS) Substation

Many applications of superconducting technology such as HTS cable, HTS transformer, HTS fault current limiter (FCL), superconducting magnetic energy storage (SMES) are analyzed. HTS cable, HTS FCL and SMES are commercially available now but their installations are still limited. HTS transformer is expected to go into market in a few years. The proposed applications are essential equipment in a substation. A distributed superconducting substation is feasible: Superconducting substation contains HTS transformer as the main transformer, HTS cable for conducting, superconducting fault current limiter (SFCL) for fault current limiting and SMES for controlling voltage stability and power quality problems. The substations will have one cryogenic refrigerator system to provide liquid helium for every HTS device. This would be more economical than using one cryogenic refrigerator for each HTS device. The superconducting substation meets the requirement for a green field substation with the respect of high efficiency, reliability, flexibility, reduced CO_2 emission, aesthetic view and safety.

A HTS system integrated concept is shown in Fig. 8.13. The substation has capacity of 100 MV·A at 24 kV and substation area is 60 m × 40 m.

The specifications and cryogenic system cost for each HTS device are presented in Tab. 8.6. The price of cryogenic system increases with capacity required. Current available large

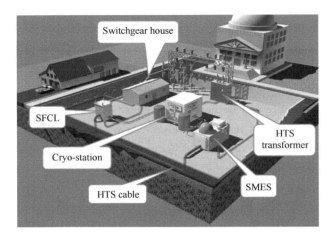

Fig. 8.13 A HTS system integrated concept

scale cryocooler device has average price of $100-150/W. In this example, the price $150/W is selected. The total cost may not be exact due to the cost of additional cryogenic lines connecting cryocooler and devices. The costs of devices are estimated based on the percent of cryogenic system cost over the total HTS cost.

Tab. 8.6 Specifications and cryogenic system cost

Component	Specification	Cryo-power/kW	Cryo Cost ($)
Transformer	100 MV·A, three-phase	240	450000
SFCL	30 MV·A	240	450000
SMES	30 MV·A	12	50000
Transmission line	100 MV·A, three-phase, 500 m	260	500000
Total-separate refrigerators		752	1450000
Super single refrigerator		700	1000000

From the above Tab. 8.6, a single refrigerator can save up to 30% and is more efficient compared to separate cryogenic generators. The superconducting substation offers many advantages over conventional substation: high transmission and distribution efficiency; high reliability, quality and flexibility; extended lifetime and reduced maintenance since the system is not affected by outside environment; high safety level, so it can be located closer to load areas; smaller size with 50% to 70% size reduction; indirectly reduced CO_2 emission and global warming; better aesthetic view.

Based on the mentioned characteristics, the superconducting substation may be a viable choice for green field substation design. In the next decade, individual HTS devices may be utilized where special features that conventional systems cannot provide are required. Due to the high cost of HTS wire and limitation in cryogenic generators and superconducting applications will not have significant impact on the utility system in the coming years.

In the long run, the future of superconductors is very bright and it promises a serious influence in the substation design. Renewable resources which are located far away from load area can be connected by superconducting cable with virtually no loss. The distribution substation will be brought closer to the load center without worrying about the effect of magnetic field and aesthetic view. This prospect will save our atmosphere from CO_2 and toxic gas and prevent the global warming, which is a worldwide concern.

(2) Solid State Transformer

Solid state transformer (SST) has the same function of stepping-up or stepping-down voltage levels as conventional iron-core transformer. The new transformer does not face the undesired properties of the conventional one such as bulky size, regular maintenance and power quality issues. High frequency converter, the heart of solid state transformer is now feasible due to silicon carbide (SiC) materials. The main advantages of SST are reduced size and weight as it uses high-frequency converter as seen in Fig. 8.14.

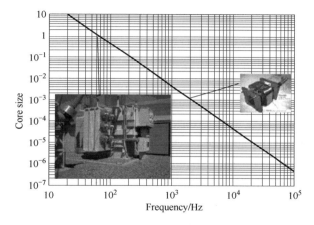

Fig. 8.14 Superconducting substation

Solid state transformer has a big potential in replacing conventional transformer in transmission and distribution substations.

Fig. 8.15 shows the diagrams of conventional and solid state transformer based substations. Due to the operation of semiconductors, once it has ceased to operate, no power will pass the high frequency transformer. Hence, HF transformer also acts as a circuit breaker. There will be no circuit breaker needed before and after the transformer like in traditional design. The substation area is significantly reduced thanks to the smaller size of SST (about 75%) and lack of circuit breakers.

The most significance that solid state transformer brings to the proposed design is the degree of controllability in transforming AC voltage. For example, phase balancing and harmonic distortion are inherently regulated since power is converted though a single HF converter.

Another feature is that either side of SST can operate asynchronously so it can have the same functionality as back-to-back HVDC or variable frequency transformer. The waveforms of either side of SST can be AC or DC, thus a DC converter system can provide interfaces to batteries for energy

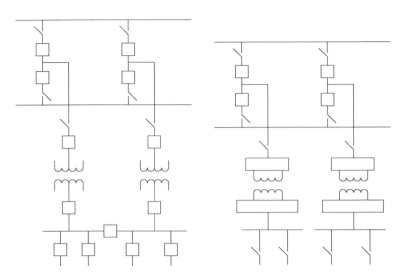

Fig. 8.15 Simplified conventional transformer substation (left) and solid state transformer substation (right).

storage and connection to fuel cells or solar cells. This ability can be considered a potential form of HVDC terminal with the integration of transformer, especially when renewable energy becomes one of the main energy sources.

The solid state transformer with silicon carbide material has many advantages over conventional transformer in operation and size issues. The properties and features of SiC SST are inherently essential for a smart gird that the national electric network is turning to. The commercial availability of SST is hard to identify because most SST projects are under research and no prototype of SST has been made. Another factor that hinders the development of SST is its high cost, at least 20 times more than the conventional transformer. SST with high frequency converter is also a highly potential alternative to a conventional HVDC station. It provides benefit in dynamic performance, rigid space requirement, reduced filtering requirement etc.

SiC technology definitely will take an important part in the advancement of power electronics in transmission and distribution systems. High voltage power electronics devices will have higher efficiency, less complexity, smaller size at affordable cost, challenging the conventional AC devices. However, the use of SST and other solid state devices still need further development to be commercially available and viable.

3. Secondary Equipment Design

Fiber-Optic Multiplexed Sensors and Control Networks in the Future Substation

An idea to multiplex data from multiple sensors on the digital communication link and then use the data at the substation level by different processing units were initiated some time ago. Recent developments in the standards for substation automation integration are allowing interconnection among IEDs from different vendors available in modern substations into one system.

In this section, a multiplexed sensor network is introduced for bringing signals into a control house very efficiently. A common signal processing set of feature extractors that will serve multiple applications in the substation is placed at the point where the analog to digital conversion takes place. The integration offers the flexibility in defining new applications that can be made transparent to the given substation layout or sensor network arrangement due to the availability of all data in the same format and at the same location (database).

The Fabry-Perot interferometer (FPI), also called the Fabry-Perot etalon, consists of two mirrors of reflectance R_1 and R_2 separated by a cavity of length L. It is used as a sensor that allows multiplexing of analog measurements.

Benefits of the FPI over conventional sensing technologies for instrumentation of the electric power grid include:

1) Immunity to electromagnetic interference, reduced susceptibility to lightning damage, and freedom from grounding problems, which affect other sensors in the presence of high electrical currents and voltages.

2) The ability to locate electronic equipment used in sensor monitoring and signal processing at remote distances from the sensing elements themselves.

3) High sensitivity to a variety of measurands.

4) The ability to multiplex many sensors to diverse types over a single optical fiber lead connection.

5) Small size and light weight for the sensing elements.

6) The potential for reduced life-cycle cost of instrumenting the electrical power grid.

Multiplexing is defined as the use of one optical source to supply light to multiple sensors, the use of one photodetector to convert the optical signal from multiple sensors, and the use of one electronic signal processor to compute measurand values for multiple sensors. Multiplexing reduces the cost per sensor. Its application is essential to cost-effective instrumentation of substations, where many points are to be remotely monitored. Architecture of the multiplexed sensor network, together with an associated signal conditioning unit (SCU) is shown in Fig. 8.16.

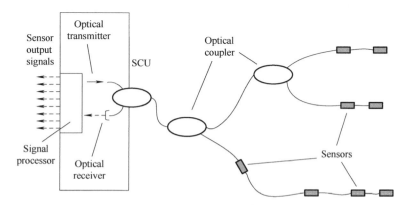

Fig. 8.16 Multiplexing arrangements for FFPI sensors

This distributed processing paradigm represents a conceptual shift from the conventional centralized model, in which all sensor output are sent to the central location for processing and decision making, saving precious transmission bandwidth and computing power.

4. Intelligent Economic Alarm Processor (IEAP) Concept and Design

The advent of electricity market deregulation has placed great emphasis on the availability of information, the analysis of this information, and the subsequent decision-making to optimize system operation in a competitive environment. This creates a need for better ways of correlating the market activity with the physical system states in real time and sharing such information among market participants. Future "intelligent" substations should play an important role in the overall "smart grid" by providing such information.

Since the power system events affect not only the operation of the power system, but also the electricity market, it can be conjectured that the importance of an electric event should be expressed in terms of the economic importance, and the economic impact should be correlated with electrical alarms. Therefore, it is proposed that alarm issuance and alarm processing should include economic information in addition to the traditional alarms.

In this section, intelligent economic alarm processor (IEAP) architecture to bring the electricity market function into the future substation design is proposed. The basic concept is to link the electricity market operation with real-time monitoring of the physical grid providing market participants and operators with economic information associated with trends in the physical system. The alarms are ranked based on the economic severity. In the proposed approach, a set of predetermined events that would give certain suppliers the ability to exercise market power will trigger an alarm. The new alarm processor proposed in this study further extends that original idea. It first gives a list of the fault occurrence possibilities based on the SCADA/IED signals received. Following these events, changes in power flows, LMPs and other economic indices is calculated and analyzed. A closer cause and effect relationship between the physical power system and the market is provided. Both physical and economic alarms are translated into easy-to-understand information to operators and market participants.

(1) Basic Assumptions

The market structure for buying, selling and scheduling electricity includes forward bilateral contracts as well as centrally coordinated markets for day ahead, hour ahead and real-time energy and ancillary services. Once the forward markets have closed, the real-time market operation coincides with real-time system operations. Schedules from the forward markets are implemented in the real-time dispatch and resources made available through the markets to provide ancillary services are selected and dispatched by the system operator for balancing (or load following) and regulation. In our example the LMPs from the real-time market are used for financial settlements of the real-time dispatch and transactions.

When an operating parameter, such as voltage, exceeds acceptable threshold, the system shifts spontaneously (dotted line in Fig. 8.17) to an unstable "emergency" state. The result is usually an automatic control action (solid line), such as the tripping of a relay, which takes the system into a more stable but not fully functional "restorative" state.

Analogous states and transitions are also applicable in power markets, with some notable differences, as shown in Fig. 8.17.

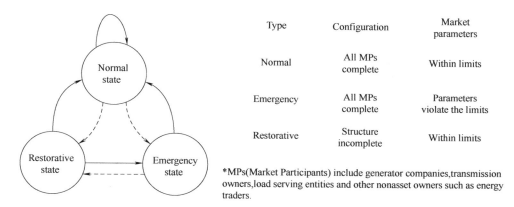

Fig. 8.17 Grid and market operating states

If system reliability is not immediately threatened, the intelligent economic alarm processor, proposed below, would give market participants advanced notice of an imminent need to find additional resources to serve scheduled loads, find replacement transmission transfer capability, or meet ancillary service needs. Marketers may often be able to find economic resources more readily if they are given advanced notice about the physical state of the system.

(2) Intelligent Economic Alarm Processor (IEAP) model

The overall architecture of the proposed IEAP model is shown in Fig. 8.18.

The fault analysis module uses a fuzzy-reasoning petri-nets (FRPN) alarm diagnosis model which has been proposed in our previous work. This solution has following characteristics:

1) Possessing the strength of both expert system and fuzzy logic as well as parallel information processing.

2) Providing the optimal design of the structure of FRPN diagnosis model.

3) Giving an effective matrix based reasoning execution algorithm.

4) Having cause-effect relationship of the fault.

5) Anomalous changes in the LMPs.

6) Triggering alarms based on power transfer volumes (as with the MW triggers for generation or lines in conventional alarms).

7) Identifying predicated limitations of available transmission capability (ATC) that is problematic.

8) Identifying energy needs as a consequence of planned events.

9) Predicting high reactive power demands.

5. Benefits of the Future Substation Design

In this part, a vision about future substation design of 20 years, 50 years or even more has been proposed. In a near future, there is no feasible technology that can replace AIS or GIS totally. The vision of GIS will keep changing to meet the criteria of green field substation more fully. Some of desirable changes in GIS technology will appear in the near future. The appearance of fault current limiter will reduce the number of circuit breakers and short circuit current to clear. Thus, a simpler breaker scheme will lead to lower cost. Solid state breaker if available could eliminate the mechanical drive and simplify the geometry so that GIS could be designed in a much simpler and cost-effective way.

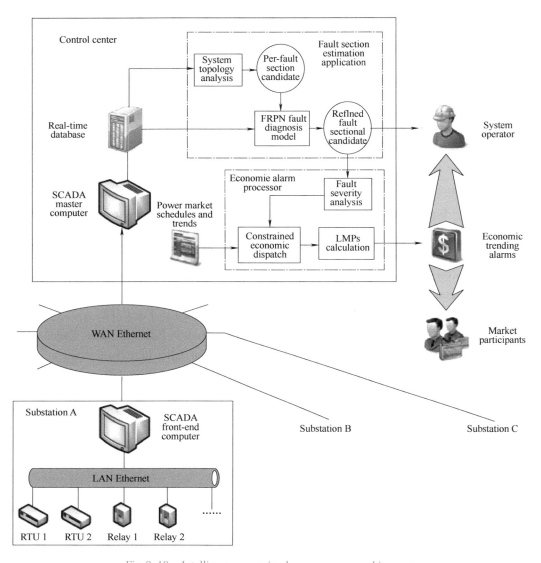

Fig. 8.18 Intelligent economic alarm processor architecture

A distributed superconducting substation is feasible. Superconducting substation containing HTS transformer as the main transformer, HTS cable for conducting, SFCL for fault current limiter and SMES for voltage stability and quality problem are envisioned. The substation uses one cryogenic refrigerator system to provide liquid helium for every HTS device. This would be more economical than using one cryogenic refrigerator for each HTS device. With the new generation of superconducting cables, the power flow is increased 2 to 3 times from that of the existing right of way. Economic losses from outage or quality disturbance are rare. Importantly, the environmental impacts are reduced significantly. HTS substation is expected to come to market in 20 to 30 years.

SiC technology definitely will take an important part in the advancement of power electronics in transmission and distribution systems. High voltage power electronics devices will have higher efficiency, less complexity, smaller size at affordable cost, challenging the conventional AC devices. Superconducting substation will be able to deliver large amount of energy over a long distance into load area. To be commercial available SST and other solid state devices still need further development. The proposed economic alarm processor would send signals changes including the LMPs, congestions, shadow prices etc. to all the market participants, which will allow them to know information at a variety of levels needed to:

1) Access the short term transmission needs in the system;

2) Allow for operators to redispatch generators based on scheduled transactions and real time market needs;

3) Make the power market more transparent by providing information to all participants;

4) Assist in making transmission operating decisions optimal for economic efficiency as well as for system reliability;

5) Allow market participants to identify trends in LMP, line loading and demand levels in order to make transactions in anticipation of these trends.

8.3 The Smart Grid

8.3.1 Introduction

"Smart grid" is one of the major trends and markets which involves the whole energy conversion chain from generation to consumer. The power flow will change from a unidirectional power flow (from centralized generation via the transmission grids and distribution grids to the customers) to a bidirectional power flow. Furthermore, the way a power system is operated changes from the hierarchical top-down approach to a distributed control.

One of the main points about smart grid is an increased level of observability and controllability of a complex power system. This can only be achieved by an increased level of information sharing between the individual components and sub-systems of the power system. Standardization plays a key role in providing the ability of information sharing which will be required to enable the development of new applications for a future power system.

1. Smart Grid Drivers

Efficient and reliable transmission and distribution of electricity is a fundamental requirement for providing societies and economics with essential energy resources. The utilities in the industrialized countries are today in a period of change and agitation. On one hand large parts of the power grid infrastructure are reaching their designed end of life time, since a large portion of the equipment was installed in the 1960s. On the other hand there is a strong political and regulatory push for more competition and lower energy prices, more energy efficiency and an increased use of renewable energy like solar, wind, biomasses and water.

In industrialized countries the load demand has decreased or remained constant in the previous decade, whereas developing countries have shown a rapidly increasing load demand. Aging equipment, dispersed generation as well as load increase might lead to highly utilized equipment during peak load conditions. If the upgrade of the power grid should be reduced to a minimum, new ways of operating power systems have be found and established. In many countries regulators and liberalization are forcing utilities to reduce costs for the transmission and distribution of electrical energy. Therefore new methods (mainly based on the efforts of modern information and communication techniques) to operate power systems are required to secure a sustainable, secure and competitive energy supply.

The key market drivers behind smart grid solutions are:
1) Need for more energy;
2) Increased usage of renewable energy resources;
3) Sustainability;
4) Competitive energy prices;
5) Security of supply;
6) Ageing infrastructure and workforce.

The utilities have to master the following challenges:
1) High power system loading;
2) Increasing distance between generation and load;
3) Fluctuating renewables;
4) New loads (hybrid/e-cars);
5) Increased use of distributed energy resources;
6) Cost pressure;
7) Utility unbundling;
8) Increased energy trading;
9) Transparent consumption & pricing for the consumer;
10) Significant regulatory push.

The priority of local drivers and challenges might differ from place to place.

2. **Some Examples**

China is promoting the development of smart grid because of the high load increase and the need to integrate renewable energy sources.

The Indian power system is characterized by high inefficiency because of high losses (technical as well as very high non-technical losses). Smart metering and flexible power system operation will make a change for the better.

In all countries with high portion of overhead lines in the distribution grid the frequency of outages is high. The number of outages, outage duration and energy not delivered in time can be reduced by using smart grid technologies.

8.3.2 Smart Grid Definitions

"Smart grid" is today used as marketing term, rather than a technical definition. For this reason there is no well defined and commonly accepted scope of what "smart" is and what it is not. However smart technologies improve the observability and/or the controllability of the power system.

Thereby smart grid technologies help to convert the power grid from a static infrastructure to be operated as designed, to a flexible, "living" infrastructure operated proactively. SG3 defines smart grid as the concept of modernizing the electric grid. The smart grid is integrating the electrical and information technologies in between any point of generation and any point of consumption.

Examples:

1) Smart metering could significantly improve knowledge of what is happening in the distribution grid, which nowadays is operated rather blindly. For the transmission grid an improvement of the observability of system-wide dynamic phenomena is achieved by wide area monitoring and system integrity protection schemes.

2) HVDC and FACTs improve the controllability of the transmission grid. Both are actuators, e.g. to control the power flow. The controllability of the distribution grid is improved by load control and automated distribution switches.

3) Common to most of the smart grid technologies is an increased use of communication and IT technologies, including an increased interaction and integration of formerly separated systems.

The European Technology Platform Smart Grid (ETPSG) defines smart grid as follows:

A smart grid is an electricity network that can intelligently integrate the actions of all users connected to it—generators, consumers and those that do both—in order to efficiently deliver sustainable, economic and secure electricity supplies.

A smart grid employs innovative products and services together with intelligent monitoring, control, communication, and self-healing technologies to:

1) Better facilitate the connection and operation of generators of all sizes and technologies;
2) Allow consumers to play a part in optimizing the operation of the system;
3) Provide consumers with greater information and choice of supply;
4) Significantly reduce the environmental impact of the whole electricity supply system;
5) Deliver enhanced levels of reliability and security of supply.

Smart grid deployment must include not only technology, market and commercial considerations, environmental impact, regulatory framework, standardization usage, ICT (information & communication technology) and migration strategy but also societal requirements and governmental edicts.

8.3.3 Smart Grid Components

Smart Grid is the combination of subsets of the following elements into an integrated solution meeting the business objectives of the major players, i.e. a smart grid solution needs to be tailored to the users' needs (see Fig. 8.19).

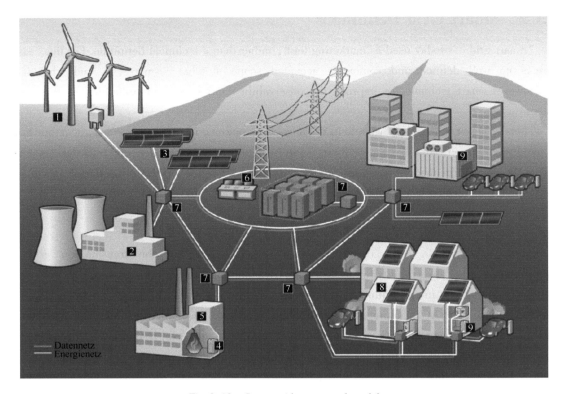

Fig. 8.19 Smart grid conceptual model

The smart grid consists of the following:

1. **Customer/Prosumer**

Smart consumption will enable demand response and lies at the interface between distribution management and building automation.

Local production is currently not a large component, however it is proposed as a future driver of smart grid requirements.

Smart homes are houses which are equipped with a home automation system that automate and enhance living. A home automation system interconnects a variety of control products for lighting, shutters and blinds, HVAC, appliances and other devices with a common network infrastructure to enable energy efficient, economical and reliable operation of homes with increased comfort.

Building automation and control system (BACS) is the brain of the building. BACS includes the instrumentation, control and management technology for all building structures, plant, outdoor facilities and other equipment capable of automation. BACS consists of all the products and services required for automatic control including logic functions, controls, monitoring, optimization, operation, manual intervention and management, for the energy-efficient, economical and reliable operation of buildings.

1) Efficient direct current transmission with HVDC Light enables energy generation in large

offshore wind farms.

2) Fossil power plants work efficiently benefits from the improvement efficiency and are only needed during off-peak periods.

3) In sunny regions, a lot of energy is generated in solar farms.

4) The decentralised combined heat and power plant efficiently supplies industrial enterprises as well as residential and functional buildings with energy, and excess energy is fed into the grid.

5) Through industrial and process automation, the industrial company works efficiently and productively. Energy management and connected intelligent devices make it a smart production facility.

6) Efficient transmission and distribution grids ensure low-loss electricity transport, even over long distances.

7) Computers process the data from the electronic meters and control energy producers and consumers. If possible, the control logic ensures the balance between electricity feed-in and withdrawal on site: in a street, in the local grid or in the distribution grid.

8) Smart homes are efficient and intelligent. Smart meters and comprehensive building system technology integrate the consumer actively. Electric cars serve as a power buffer. If more energy is generated in the decentralised plants than required, it is fed into the grid. The consumer who has become a producer receives a remuneration for this.

9) Building automation also makes functional buildings efficient. With smart meters and building system technology based on KNX, they become intelligent buildings.

2. Bulk Generation

Smart generation will include the increased use of power electronics in order to control harmonics, fault ride-through and fluctuating generation from renewables as well as the required increased flexibility of conventional fossil power plants due to the increased fluctuation of feed from the renewables.

3. Power Grid (Transmission and Distribution)

Substation automation & protection is the backbone for a secure transmission grid operation. During recent years serial bus communication has been introduced (IEC 61850). Security is based on protection schemes.

Power quality and power monitoring systems act in a very similar way to quality management systems in companies. They are independent from operation, control and management systems and supervise all activities and assets/electrical equipments in a corresponding grid. Therefore such systems can be used as "early warning systems" and are a must to analyze faults and to find out the corresponding reasons.

The energy management system (EMS) is the control centre for the transmission grid. Today customers require an open architecture to enable an easy IT integration and a better support to avoid blackouts (e. g. phasor measurements, visualization of the grid status, dynamic network stability analysis).

In contrast to traditional protection devices, which protect the primary equipment (e.g. transformers) from fatal fault currents, the decision support systems (DSS) and system integrity protection schemes (SIPS) protect the power systems from instabilities and blackouts. System integrity protection schemes (SIPS) will enhance the target of protection devices, to protect the primary equipment (e.g. transformers) from fatal fault currents in such a way that uncontrollable chain reactions, initiated by protective actions, are avoided by limited load shedding actions.

Power electronics is among the "actuators" in the power grid. Systems like HVDC and FACTs enable actual control of the power flow and can help to increase transport capacity without increasing short circuit power.

Asset management systems and condition monitoring devices are promising tools to optimize the OPEX and CAPEX spending of utilities. Condition-based maintenance, for example, allows the reduction of maintenance costs without sacrificing reliability. Furthermore they may also be used to utilize additional transport capacity due to better cooling of primary equipment, e.g. transmission lines on winter days.

Distribution automation and protection: Whereas automated operation and remote control is state of the art for the transmission grid, mass deployment of distribution automation is only recently becoming more frequent, leading to "smart gears". Countries like the United States of America, where overhead lines are frequently used, benefit most. Advanced distribution automation concepts promote automatic self-configuration features, reducing outage times to a minimum (self-healing grids). Another step further is the use of distributed energy resources to create self-contained cells (microgrids). Microgrids can help to assure energy supply in distribution grids even when the transmission grid has a blackout.

The distribution management system (DMS) is the counterpart to the EMS and is therefore the control center for the distribution grid. In countries where outages are a frequent problem, the outage management system (OMS) is an important component of the DMS. Other important components are fault location and interfaces to geographic information systems (GIS).

Smart meter is a generic term for electronic meters with a communication link. Advanced metering infrastructure (AMI) allows remote meter configuration, dynamic tariffs, power quality monitoring and load control. Advanced systems integrate the metering infrastructure with distribution automation.

4. Communication

Communication as a whole is the backbone of smart grid. Only by exchanging information on a syntactic and semantic level can the benefits of smart grid be achieved.

Security of a critical infrastructure has always been an issue. However smart grid solutions will see an enormous increase in the exchange of data both for observability and for controllability. Therefore security of this data exchange and the physical components behind it will have an increased impact.

8.3.4 Smart Grid in China

In both China and the United States, the power sector is a dominant source of fossil-fuel combustion and greenhouse gas emissions. Decarbonizing the power sector by both reducing the use of fossil fuels (particularly coal) and using them more efficiently, as well as integrating greater amounts of renewable, low-carbon energy sources like wind and solar, is an essential measure for reducing anthropogenic greenhouse gas emissions.

In order to do this, both countries will need to develop and build an updated electricity transmission infrastructure—a so-called smart grid—that can serve as a more sophisticated and intuitive framework for transmitting and using electricity more efficiently. But what exactly constitutes a smart grid, and what are the concrete mechanisms by which it would reduce greenhouse gas (GHG) emissions? The purpose of this section is to answer those questions.

It is important to note here, however, that the concept of smart grid in most of the world, including the United States, differs significantly from the term strong, smart grid being promoted in China. In most of the world, smart grid refers only to the communications and information technologies that will enhance the use and delivery of electricity. In China, the emphasis on a strong grid reflects China's focus on developing and maintaining a stable power supply and includes its plans to build thousands of kilometers of ultra-high voltage (UHV) transmission lines in the next 10 years to connect the coal, wind, hydro and solar-rich west with the high-demand east. Because the strong concept is an important part of China's plans, it is discussed in this sector along with the commonly accepted smart grid concepts. Regional integration of grids will reduce the impact of intermittent wind and solar resources, yielding more reliable generation, and will also match power supply with power demand.

As a primer on the concept of a strong, smart grid, this sector will explain the possible carbon reduction mechanisms that are enabled or enhanced by strong, smart grid technologies:

1) Section 1 begins by defining the essential elements of a smart grid.

2) Section 2 reviews and explains the carbon reduction mechanisms covered in two important reports published by the United States Electric Power Research Institute (EPRI) and State Grid Corporation of China (SGCC)—the larger of China's two main grid companies—that provide estimates of the GHG emissions enabled by smart grid in the United States and China, respectively.

1. What is a (Strong) Smart Grid

A strong, smart grid is an interconnected system of information and communication technologies and electricity generation, transmission, distribution, and end-use technologies that will:

1) Enable consumers to manage their power usage and choose the most economically efficient products and services (two-way communications);

2) Maintain electricity delivery system reliability and stability enhanced by automation (intelligent monitoring and control);

3) Integrate the most environmentally benign generation alternatives including renewable resources and energy storage (two-way power transmission, renewable integration).

It is important to note here that the term smart grid does not require the inclusion of all of these components, and may be of even narrower scope than the descriptions herein. Actual smart grid projects may involve only one of the following components:

(1) Two-Way Communications

The smart grid enables real-time, two-way communications throughout the whole electrical system, from generation to transmission to distribution to end-use consumption. This is in strong contrast to the traditional electric grid, which does not allow communication between power utilities and residential consumers, and relies on customers for reports on power outages, information that the traditional grid cannot supply.

Smart grid also allows for real-time pricing information, so that consumers can make decisions about when to use electricity. This does not mean that consumers must continuously monitor pricing information; a smart air conditioning unit can be set to automatically raise its thermostat setting when the electricity price goes above a certain threshold (this would be expected to happen at peak load times—hot summer days when most air conditioning units are running). Instead, it means that customers will no longer have to wait until billing statements in order to learn about pricing. From the utility's perspective, real-time pricing is better for business because it helps shift consumption from times of peak usage to times of lower usage. Consumption information can also be automatically sent from consumers' smart meters to utilities so that manual meter reading will no longer be necessary.

(2) Two-Way Power Transmission

Whereas traditional electric grid allows only one-way electricity flow from large-scale generators through transmission and distribution infrastructure to consumers, the smart grid enables two-way power flow so that consumers can also be generators. Distributed generation could decrease the risk of large-scale power outages by making use of distributed space for small-scale installations (e.g., rooftop solar photovoltaic systems), and placing generation closer to consumption so as to reduce transmission losses and the need for new transmission capacity.

In addition, two-way electricity flow enables plug-in electric vehicles (EVs) to act as peak-shifting energy storage. Most EV users probably charge their vehicles at night while demand is off-peak and electricity prices are low. During the day, if charging stations are widespread, vehicles can be plugged in to provide electricity to the grid when it is most needed. In this way, EV owners can actually make money by buying electricity when it is cheap and selling it back at higher prices. From the utility's perspective, not only does this system reduce peak loads, it also maintains a higher base load at night, meaning its generators can run more continuously and, thus, more efficiently.

(3) Intelligent Monitoring and Control

The traditional grid has very limited ability to monitor the status of the electrical system and react to potential problems before they arise. The smart grid, with its communications infrastructure, is able to detect potential problems before they become severe and promptly deploy measures to solve those problems.

For example, one common issue for utility companies is reactive power, which is created by some loads and generators, and congests transmission lines but does not deliver energy. Utility companies and large commercial and industrial customers have a strong incentive to optimize power factor in order to minimize reactive power. A smart grid can enhance the system's ability to sense the amount of reactive power flowing in real-time and activate equipment that provides reactive power compensation in the correct location, clearing up congestion and increasing the flow of real power or increasing the flow of reactive power to maintain voltages, as the case may be. This capability increases distribution system efficiency and enhances grid stability.

(4) **Renewable Integration**

The variable and uncertain nature of renewable energy sources like wind and solar introduces difficulties for integrating renewable energy into the grid. For example, at times there are strong winds when demand for power is very low. To avoid shutting down the turbines (currently a common practice), the grid must be able to transmit the electricity either to areas of higher load or to energy storage facilities. And, when the wind is weaker than forecasted, smart grid may use demand response to reduce loads.

2. **Estimating the GHG Emissions Reductions Enabled Through a (Strong) Smart Grid in the United States and China**

Having defined the key characteristics that distinguish a smart grid from a conventional grid, the next step is to describe the concrete mechanisms by which a strong, smart grid can save energy and reduce GHG emissions, and to estimate the possible emissions associated with each mechanism. The Electric Power Research Institute's 2008 report, *The Green Grid: Energy Savings and Carbon Emissions Reductions Enabled by a Smart Grid*, provides such as estimate for the United States, while the State Grid Corporation of China's 2010 *Green Development White Paper* provides a similar estimate for China, based on the benefits of a strong, smart grid.

There is some overlap in the mechanisms examined in each report, especially those for reducing transmission loss, but for many reasons—including the different years used in the two reports, the various mechanisms examined, and contrast in assumptions—an apples-to-apples comparison of the emissions reductions from smart grid in the United States and China is impossible.

Also, it is clear from the mechanisms listed and each mechanism's relative importance, that the two reports assume different priorities with respect to smart grid development in each country. The EPRI report, for example, attributes large energy savings to the increased feedback customers will receive on their energy usage and the accelerated deployment of energy efficiency programs, as well as savings from peak load management. The SGCC report, on the other hand, places little emphasis on the effects of energy conservation and consumer energy efficiency measures. Rather, the SGCC report assumes China has extremely aggressive plans for EV deployment, and, as evident from the gigantic savings associated with increasing the efficiency of coal-fired power generation, that China will remain heavily dependent on coal.

1. outage   *n.* 运行中断   2. intermittent   *adj.* 间歇的，断断续续的

3. grab　*vi.* 攫取；夺取
4. hydro　*n.* 水力发电；电力
5. vintage　*n.* 特定年份（或地方）酿制的酒　*adj.* 典型的；优质的
6. switchyard　*n.* 配电装置；开关站
7. proliferation　*n.* 增殖，扩散；分芽繁殖
8. vulnerability　*n.* 脆弱性；易损性；弱点
9. busbar　*n.* 母线，母线槽；汇流排
10. tremendously　*adv.* 特别；巨大，非常地
11. aesthetic　*adj.* 美的；美学的；审美的
12. auxiliary　*adj.* 辅助的；副的；附加的　*n.* 助动词；辅助者，辅助物；附属机构
13. substation　*n.* 变电所；分所；分局；分台
14. criteria　*n.* 标准，条件（criterion 的复数）
15. fiber　*n.* 光纤；纤维
16. phasor　*n.* 相量；矢量
17. scheme　*n.* 计划；体制　*vt.* 计划；策划
18. perimeter　*n.* 周界；周长；视野计
19. database　*n.* 数据库，资料库
20. cabling　*n.* 电缆；多芯导线；布线；被覆线
21. streamline　*vt.* 使合理化；使成流线型　*n.* 流线型；流线
22. versatile　*adj.* 通用的，万能的；多才多艺的
23. verification　*n.* 确认，查证；核实
24. sequence　*n.* 序列；顺序
25. topology　*n.* 拓扑学；地志学；局部解剖学
26. transient　*adj.* 短暂的；路过的　*n.* 瞬变现象
27. diagnosis　*n.* 诊断
28. conjecture　*vt.* 推测
29. cryogenic　*adj.* 低温学的；低温实验法的；冷冻的
30. helium　*n.* 氦
31. carbide　*n.* 碳化物；碳化钙
32. harmonic　*n.* 谐波；和声　*adj.* 和声的；和谐的
33. photodetector　*n.* 光电探测器
34. paradigm　*n.* 范例；示例；模范；规范
35. threshold　*n.* 极限；临界值
36. transparent　*adj.* 透明的；易懂的；显然的
37. hierarchical　*adj.* 分层的
38. agitation　*n.* 搅动；纷乱；激动
39. fluctuate　*vi.* 波动；涨落；动摇
40. backbone　*n.* 主干网；支柱；决心
41. semantic　*adj.* 语义的；语义学的
42. syntactic　*adj.* 句法的；语法的
43. decarbonize　*vt.* 除去碳素；脱去…的碳
44. promptly　*adv.* 迅速地；立即地；敏捷地

NOTES

1. It is also designed to link the grid to large scale solar and wind projects that are built far away from the cities and suburbs where people need electricity.

智能电网也被设计用来对建在远离城市和郊区的大规模太阳能发电站和风电场进行联网，以满足（城市和郊区的）人们的电力需求。

2. When building a new substation, which does not happen very often in the USA, one has an opportunity to use prior experiences when deciding on the requirements of the new design.

虽然这在美国并不常见，但当建立一个新变电站时，有责任利用以前的经验对是否需要进行新的设计做决定。

3. As may be noted, the GIS designs are supposed to be applied where one or more of the

following features are desirable: limited space, extreme environmental conditions, required low environmental impact and less maintenance.

值得注意的是,气体绝缘开关的设计应该适用于具备以下一个或多个特征的地方:有限的空间,极端的环境条件,要求对环境影响低和较少的维护。

4. It takes various relay reports and files as inputs and using embedded expert system generates a report on the results of analysis.

它以各种继电保护报告和文件作为输入,然后使用嵌入式专家系统生成一个结果分析报告。

5. A critical implication of this restructuring will be to make electricity markets even more intensely data driven, creating a need for better ways of monitoring market activity in real-time and sharing information among market participants.

这一重组的关键意义就是让更密集的数据驱动电力市场,为实时监管市场活动提供更好的办法,实现市场参与者之间的信息共享。

6. Due to the operation of semiconductors, once it has ceased to operate, no power will pass the high frequency transformer. Hence, HF transformer also acts as a circuit breaker.

由于半导体的运作(受控运行),一旦它停止运行,电能就不能通过高频变压器。因此,高频变压器也可作为一个断路器使用。

7. An idea to multiplex data from multiple sensors on the digital communication link and then use the data at the substation level by different processing units were initiated some time ago.

在数字通信链路上复用多个传感器的数据,然后通过不同的处理单元在变电站层级进行使用,这种想法是在一段时间以前产生的。

8. This distributed processing paradigm represents a conceptual shift from the conventional centralized model, in which all sensor output are sent to the central location for processing and decision making, saving precious transmission bandwidth and computing power.

在传统集中式处理模式中,所有传感器输出被发送到中央位置进行处理和决策。这种分布式处理模式是从传统集中式模式演变而来,可节省宝贵的传输带宽和计算能力。

9. Solid state breaker if available could eliminate the mechanical drive and simplify the geometry so that GIS could be designed in a much simpler and cost-effective way.

如果条件允许,固态断路器可以消除机械动作并简化结构,从而使 GIS 设计更简单,更经济。

10. Smart metering could significantly improve knowledge of what is happening in the distribution grid, which nowadays is operated rather blindly.

智能电表可以显著提高对配电网中发生的事情的掌握程度,而不是像现在对配电网几乎是盲操作。

11. To avoid shutting down the turbines (currently a common practice), the grid must be able to transmit the electricity either to areas of higher load or to energy storage facilities.

为了避免关闭汽轮机(目前一种普遍的做法),电网必须有能力将电力传输到负荷更高的地区或储能装置中去。

Chapter 9
Wide Area Measurement Protection System

9.1 Introduction

Phasors are basic tools of AC circuit analysis, usually introduced as a means of representing steady-state sinusoidal waveforms of fundamental power frequency. Even when a power system is not quite in a steady-state, phasors are often useful in describing the behavior of the power system. For example, when the power system is undergoing electromechanical oscillations during power swings, the waveforms of voltages and currents are not in steady-state, and neither is the frequency of the power system at its nominal value. Under these conditions, as the variations of the voltages and currents are relatively slow, phasors may still be used to describe the performance of the network, the variations being treated as a series of steady-state conditions.

Recently, it has been recognized that phasors are applicable even when the waveforms are changing rather rapidly and contain significant amounts of transient components. Thus, in relaying applications, one speaks of a phasor of a voltage or current over a half-or one-cycle observation window, and calculations made with this concept of a phasor are found to agree well with the normally understood meaning of a steady-state phasor.

Measuring systems using digital computers were introduced in the power industry many years ago. However, the recent advent of computer relaying has led to new insights into the measurement process, its limitations, and its potentialities. The importance of the phase angle (contained in the phasor representation) in electric power engineering need hardly be emphasized. Recent developments in time synchronizing techniques, coupled with the computer-based measurement technique, have provided a novel opportunity to measure the phasors, and phase angle differences in real-time. This chapter reviews this technology and describes many of the research projects concerned with applications of synchronized phasor measurements.

9.2 Phasor Measurement Unit (PMU) and Its Application

9.2.1 Defination of Phasor

Consider the steady-state waveform of a nominal power frequency signal as shown in Fig. 9.1.

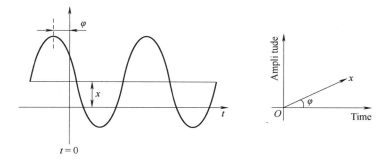

Fig. 9.1 Phasor representation of a sinusoidal

If we start our observation of this waveform at the instant $t=0$, the steady-state waveform may be represented by a complex number with a magnitude equal to the RMS value of the signal and with a phase angle equal to the angle φ. In a digital measuring system, samples of the waveform for one (nominal) period are collected, starting at $t=0$, and then the fundamental frequency component of the discrete fourier transform (DFT) is calculated according to the relation:

$$X = \frac{2}{N}\sum_{k=1}^{N} X_k e^{-j2\pi k/N}$$

where N is the total number of samples in one period, X is the phasor, and X_k is the waveform samples. This definition of the phasor has the merit that it uses a number of samples (N) of the waveform, and is the correct representation of the fundamental frequency component, when other transient components are present. When the input signal frequency is different from the nominal frequency, an error is introduced in the magnitude and the phase angle of the phasor. However, the error itself is quite interesting, and as we will see later, it can be used to determine the frequency of the input signal.

It is essential that the input waveform be filtered to eliminate aliasing errors in the DFT calculation described. Thus, the signal must not contain frequencies above the Nyquist rate: $N\Omega_0/2$. Often the input signal contains frequencies that are not harmonics of the fundamental frequency. In such cases, the phasor calculation is once again in error. One could treat the extraneous nonharmonic components as a noise signal, and then the computed phasor has an uncertainty associated with it, which can be depicted as shown in Fig. 9.2.

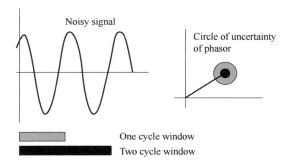

Fig. 9.2 Data window and uncertainty in phasor estimation

The circle of uncertainty is inversely proportional to the square root of the data window. More data used in computing the phasor is beneficial in this regard. It should be noted that the waveform is sampled by the measurement system continuously, and each time a new sample is acquired, a new phasor is obtained with a data window including the latest sample. The most efficient method of dealing with continuous monitoring of the input waveforms is to use a recursive form of the phasor equation.

Phasors can be measured for each of the three phases, and the positive sequence phasor computed according to its definition:

$$X_1 = \frac{1}{3}(X_a + \alpha X_b + \alpha^2 X_c)$$

where $\alpha = e^{j2\pi/3}$.

In the context of power system performance evaluation, the positive sequence voltages and currents are far more useful than the corresponding phase quantities.

When several voltages and currents in a power system are measured and converted to phasors in this fashion, they are on a common reference if they are sampled at precisely the same instant. This is easy to achieve in a substation, where the common sampling clock pulses can be distributed to all the measuring systems. However, to measure common-reference phasors in substations separated from each other by long distances, the task of synchronizing the sampling clocks is not a trivial one. Over the years, recognizing the importance of phasors and phase angle difference measurements between remote points of a system, many attempts have been made to synchronize the phasor measurements. None of these early attempts were too successful, as the technology of the earlier era is put severe limitations on what could be accomplished. It is only in recent years that the technology has reached a stage, whereby we can synchronize the sampling processes in distant substations economically, and with an error of less than 1 microsecond.

A 1-microsecond error translates into 0.021" for a 60 Hz system and is certainly more accurate than any presently conceived application would demand.

9.2.2 Sources of Synchronization

Synchronization signals could be distributed over any of the traditional communication media currently in use in power systems. Most communication systems, such as leased lines, microwave, or AM radio broadcasts, place a limit on the achievable accuracy of synchronization, which is too coarse to be of practical use. Fiberoptic links, where available, could be used to provide high precision synchronization signals, if a dedicated fiber is available for this purpose. If a multiplexed fiber channel is used, synchronization errors of the order of 100 microseconds are possible, and are not acceptable for power system measurements. GOES satellite systems have also been used for synchronization purposes, but their performance is not sufficiently accurate.

The technique of choice at present is the Navstar GPS satellite transmissions. This system is designed primarily for navigational purposes, but it furnishes a common-access timing pulse, which is accurate to within 1 microsecond at any location on earth. The system uses transmissions from a constellation of satellites in nonstationary orbits at about 10000 miles above the earth's surface. For accurate acquisition of the timing pulse, only one of the satellites need be visible to the antenna. The antenna is small (about the size of a water pitcher), and can be easily mounted on the roof of a substation control house. The experience with the availability and dependability of the GPS satellite transmissions has been exceptionally good.

9.2.3 Phasor Measuring Units

The phasor measuring units (PMUs) is a device for synchronized measurement of AC voltages and currents, with a common time (angle) reference. The most common time reference is the GPS signal, which has a precision down to 1 s. In this way, the AC quantities can be time-stamped and measured as complex values and represented by their magnitude and phase angle.

PMUs using synchronization signals from the GPS satellite system have evolved into mature tools and are now being manufactured commercially. The functional block diagram of a typical PMU is shown in Fig. 9.3.

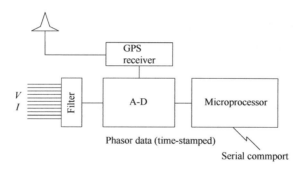

Fig. 9.3 a functional block diagram of a typical PMU

The GPS receiver provides the 1 pulse-per-second (PPS) signal, and a time tag, which consists of the year, day, hour, minute, and second. The time could be the local time, or the universal time coordinated (UTC). The 1-PPS signal is usually divided by a phase-locked oscillator into the required number of pulses per second for sampling of the analog signals. In most systems being used at present, this is 12 times per cycle of the fundamental frequency. The analog signals are derived from the voltage and current transformer secondaries, with appropriate anti-aliasing and surge filtering.

The complete unit meets the IEEE C37.90.1—1989 IEEE Standard Surge Withstand Capability (SWC) Tests for Protective Relays and Relay Systems. The microprocessor determines the positive sequence phasors according to the recursive algorithm described previously, and the timing message from the GPS, along with the sample number at the beginning of a window, is assigned to the phasor as its identifying tag. The computed string of phasors, one for each of the positive sequence measurements, is assembled in a message stream to be communicated to a remote site. The messages are transmitted over a dedicated communication line through the modems. A 4800-baud communication line can support the transmission of the phasor stream at the rate of about every 2-5 cycles of the fundamental frequency, depending upon the number of positive sequence phasors being transmitted.

9.2.4 PMU Potential Applications

Synchronized phasor measurements have become a practical proposition only during the last 3 years. As such, their potential for use in power system applications has not yet been fully realized. Although prototype systems of phasor measurement units have been installed in many utilities, it should be realized that we are dealing with a subject that is a research topic at present. No doubt, as experience with these systems is gained, other applications of phasor measurements would surface. The followings are a few of the applications of the phasor measurement systems that are being developed.

1. Measuring Frequency and Magnitude of Phasors

These applications are based upon a feature of the Fourier-transform-based phasor measurement technique and do not require synchronization of the sampling processes. The recursive positive sequence phasor calculation performed with a fixed frequency sampling clock produces a complex number (the phasor), which rotates in the complex plane with an angular velocity equal to the difference between the nominal power system frequency and the prevailing actual power system frequency If the positive sequence phasor (voltage or current) is

$$X_1 = |X_1| e^{-\varphi_1}$$

then the prevailing power system frequency is given by

$$\omega = \omega_0 + \frac{d\varphi_1}{dt}$$

For example, when the power system frequency is 61 Hz, the phasor rotates one revolution per second in a counterclockwise direction, while at 59 Hz, the phasor rotates one revolution per second in the clockwise direction. This technique offers one of the most sensitive and accurate methods of power system frequency measurement. Frequency deviations of 0.001 Hz have been measured in the laboratory and in the field. Single-phase phasors may also be used to calculate frequency with equal precision, provided additional computational measures are taken to correct the effects of off-nominal operation on the single-phase phasor calculation.

High-precision magnitude calculations of voltage or current phasors can also be made by this technique. Precision of measurement is enhanced by the use of true 16 bit A-D converters, and by using longer data windows. It has been shown that the error of estimation caused by certain types of noise signals depends inversely on the square-root of the length of the data window. Thus, using a four-cycle data window reduces the error of phasor estimation by a factor of 2, as compared to a one cycle data window. As longer data windows are used to improve phasor estimation accuracy, the attenuation factor caused by off-nominal frequency operation becomes important and must be taken into account by the computation technique used.

2. State Estimation

Modern electric utility control centers use state estimators to monitor the state of the power system. The state estimator uses various measurements (such as complex powers and voltage and current magnitudes) received from different substations, and, through an iterative nonlinear estimation procedure, calculates the power system state. The state (vector) is a collection of all the positive sequence voltage phasors of the network, and, from the time the first measurement is taken to the time when the state estimate is available, several seconds or minutes may have elapsed. Because of the time skew in the data acquisition process, as well as the time it takes to converge to a state estimate, the available state vector is at best an averaged quasi-steady-state description of the power system. Consequently, the state estimators available in present-day control centers are restricted to steady-state applications only.

Now, consider the positive sequence voltages measured by the synchronized phasor measurement units. If voltages at all system substations are measured, one would have a true simultaneous measurement of the power system state. No estimation of the state vector is necessary. From a practical point of view, it is sensible to use the positive sequence currents also, which provide data redundancy. This leads to a linear estimator of the power system state, which uses both current and voltage measurements. The estimate results from the multiplication of a constant matrix by the measurement vector, and it is extremely fast.

In addition to a much simplified static state estimator, synchronized phasor measurements also provide the first real possibility of providing a dynamic state estimator. By maintaining a continuous stream of phasor data from the substations to the control center, a state vector that can follow the system dynamics can be constructed. With normal dedicated communication circuits operating at 4800 or 9600 baud, a continuous data stream of one phasor measurement every 2-5 cycles (33.3-83.33 ms) can be sustained. Considering that the usual power system dynamic phenomena fall in the range of 0-2 Hz, it is possible to observe in real-time the power system dynamic phenomena with high fidelity at the center.

Another application of directly measured dynamic phenomena is to validate power system models used in transient stability studies. For the first time in history, synchronized phasor measurements have made possible the direct observation of system oscillations following system disturbances. By trying to simulate these events, one can learn a great deal about the models of major system components, and correct them as needed until the simulations and observed phenomena match well.

3. Instability Prediction

It is well known that the threat of instability governs most aspects of modern power system operation. The design, operation, and protection of a power system is either directly or indirectly affected by the likelihood of the system going unstable in the event of a contingency. System loading limits are based upon stability considerations, as the operating speeds of the primary and back-up protection systems, and the settings of out-of-step relays. This being the case, any improvements made in determining the imminence of an unstable condition, or controlling it, has great benefits to modern interconnected networks.

Synchronized phasor measurements offer a unique opportunity of improving the response of protection and control systems to an evolving power swing. At the foundation of all possible improvements is the prospect of predicting in real-time the outcome of an evolving transient oscillation. We will consider just the instability prediction problem here, and discuss the specific protection and control aspects in the following sections.

The traditional method of stability analysis is by direct integration of the system dynamic equations. The dimensionality of the problem forces us to use various types of simplification. Even so, the computations involved are so extensive that these techniques have been confined to offline stability studies. There have been relatively few instances where stability analysis in real-time is attempted. Transient energy functions seem to offer a usable alternative for real-time stability analysis. However, there appear to be considerable theoretical difficulties in determining the appropriate energy limits. At present, the most promising technique seems to be high-speed integration of appropriately simplified models of the power system.

A striking difference between the real-time stability prediction problem and the offline analysis is that we can actually monitor the progress of the transient in real-time, thanks to the technique of synchronized phasor measurements. Consider the evolution of a transient over time as shown in Fig. 9.4.

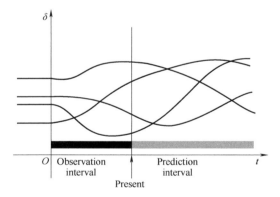

Fig. 9.4 Prediction of power system stability for adaptive out-of-step relaying

The power system itself provides us with the system trajectory up to the present time. The time track of state variables and several of their derivatives are available over an observation window, so it seems possible that the outcome of the swing for a future time interval can be calculated with relatively good, simplified models. It seems probable that predictions of up to a second in the future can be made with reasonable confidence. With this prediction capability, useful protection and control decisions can be made to change the course of the transient if that seems desirable.

4. Adaptive Relaying

One of the most obvious applications of the instability prediction technique is to provide an improved adaptive out-of-step protection function, which adapts to changing power system conditions. Out-of-step relays detect the onset of a power swing with the help of the apparent impedance seen by a distance relay. As a swing develops, the apparent impedance changes. By sensing the extreme values of the impedance change, and the time it takes to bring about the change, the outcome of the power swing (stable or unstable) is inferred (Fig. 9.5).

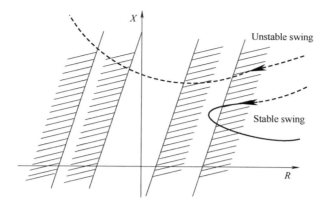

Fig. 9.5 Out-of-step relaying using impedance relays

The various settings of the out-of-step relay are determined by conducting several stability simulations for all reasonable contingencies. The problem comes in when the actual prevailing power system conditions are substantially different from those assumed in the contingencies studied. Consequently, the out-of-step relay settings may be inappropriate for the actual power system conditions.

If the outcome of a power swing can be predicted by observing the real-time data provided by the synchronized phasor measurements as outlined previously, then a more fitting response to the existing power system condition can be provided. Work along these lines is currently under way on a practical out-of-step relay to be used in the field.

These ideas are an example of the new field of adaptive relaying, which has been defined as follows: "Adaptive relaying is a protection philosophy which permits and seeks to make adjustments in various protection functions in order to make them more attuned to prevailing power conditions." Other adaptive protection possibilities, some using synchronized phasor measurements, are also currently under investigation.

5. Improved Control

Power system control elements, such as generation excitation systems, HVDC terminals, variable series capacitors, SVCs, etc., use local feedback to achieve the control objective. However, often the control objective may be defined in terms of a remote occurrence. As an example, consider the task of damping power swings between two areas by controlling (modulating) the flow on a DC line. Such a controller must have a built-in mathematical description (model), which must relate the DC power to the angle between the two regions. To the extent that the assumed model is not valid under the prevailing system conditions, the controller does not do the job for which it was intended.

With synchronized phasor measurements being brought to the controller location, it becomes possible to provide direct feed-back from the angular difference between the two systems. Studies of this nature have shown that improved control performance is achieved when a model-based controller is replaced by one based upon feed-back provided by the phasor measurement system.

6. Commercial Systems

Several electric power companies in North America and in other parts of the world are actively investigating the applications of synchronized phasor measurement units on their systems. About two dozen PMUs have been installed in the field, or are in the process of being installed. These systems were assembled from individual components, such as a GPS receiver, and an off-the-shelf microcomputer. More recently, commercial manufacture of PMUs has begun, and it can be expected that in the future all PMUs installed will be of this variety, conforming to various industry standards for substation equipment, and using the most modern signal processing technology.

The commercial systems are integrated, i.e. they include all subsystems within one chassis. As the phasor calculation is derived from samples of voltages and currents, it can be anticipated that all relays, digital fault recorders, and other measuring instruments can be designed to serve as synchronized phasor measurement units, if the GPS synchronizing signals are used to provide sampling pulses.

Anticipating that this technology is about to take off, the IEEE Power System Relaying Committee (PSRC) is developing a standard that will define the interface between the GPS receivers and the PMU, and for the communication link between the PMUs and a remote site. The standard will be based upon a paper published by a working group of the PSRC.

This emerging field has been made possible by some remarkable technological innovations in computers, signal processing, and satellite systems. From the point of view of a university professor, this has been an ideal subject with which to refute the misconception that electric power engineering is an old, unglamorous field. This type of measurement system can revolutionize the field of monitoring, protection, and control of power systems. It is with great excitement that we look for other applications, not yet thought of, that can advance the state of the art in electric power engineering.

9.3 Wide-Area Protection System

9.3.1 Introduction

This part describes basic principles and philosophy for wide-area protection schemes, also known as remedial action schemes (RAS) or system protection schemes (SPS). In the areas of power system automation and substation automation, there are two parallel trends in different directions: centralization and decentralization. More and more functions are moved from local and regional control centers toward the central or national control center. At the same time we also observe more and more "intelligence" and "decision-power" moving closer toward the actual power system process. We also see a great deal of functional integration, i.e., more and more functionality enclosed in the same hardware. This raises discussions concerning reliability (security and dependability).

There are a few basic facts and technological developments that have pushed the utility needs and the vendors' offers in wide-area protection and control.

1) The deregulated electricity market causes rather quick changes in the operational conditions. New, unknown, load flow patterns show up more frequently for the system operator.

2) Economic pressure on the electricity market and on grid operators forces them to maximize the utilization of high-voltage equipment, which often means operation closer to the limits of the system and its components. For the same reason, there is also a wish to push the limits.

3) Reliable electricity supply is continually becoming more and more essential for the society and blackouts are becoming more and more costly whenever they occur.

4) Technical developments in communication technology and measurement synchronization, e.g., for reliable voltage phasor measurements, have made the design of system wide protection solutions possible. The use of phasor measurements also provides new possibilities for state estimator functions. Their main use so far has been for wide-area measurement system (WAMS) applications.

5) There is a general trend to include both normal operation issues and disturbance handling into power system automation (PSA).

6) Wide-area disturbances during the last decades have forced/encouraged power companies to design system protection schemes to counteract voltage instability, angular instability, frequency instability, to improve damping properties or for other specific purposes, e.g., to avoid cascaded line trip.

7) A lot of research and development within universities and industry have significantly increased our knowledge about the power system phenomena causing widespread blackouts. Methods to counteract them have been or are being developed.

8) There is a heightened concern for security of power grids due to acts of coordinated sabotage, which traditionally were not considered in grid planning. Fast and efficient controls/protections are needed to stop the disruption from spreading.

Based on the situation and new technology described above, grid companies can benefit a lot. Wide-area protection and emergency control systems can be introduced in the power system, either: to increase the power system transmission capability; or to increase the power system reliability; or a combination of the two.

This means that the grid company can extend the system operational limits, i.e., operate the system closer to the physical constraints, without any investments in high voltage equipment. The operational criteria will change from "the power system should withstand the most severe credible contingency" to "the power system should withstand the most severe credible contingency, followed by protective actions from the wide-area protection system."

Certain power systems do not fulfill the design and operational criteria, which are based on the reliability requirements on the system, without installation of wide-area protection systems. The reliability requirements on such wide-area protection systems—dependability as well as security—are extremely high.

9.3.2 Requirements on Protection Compared to SCADA/EMS

Products for wide-area protection and emergency control should be designed and manufactured in a similar way as conventional equipment protection, concerning standardization, flexibility, hardware and software modularization, configuration, and functionality. Maximum benefit from available hardware that have passed different kinds of EMC and environmental tests, as well as software functions available from other protection terminals, has to be made. Standardized communication protocols and hardware are also essential.

SCADA/EMS functions based on phasor measurements and inputs from system protection terminals should in general be compatible with present and future SCADA/EMS systems.

Based on synchronized phasor measurements, more efficient state estimation can be performed. Based on fast and reliable state estimation a variety of system stability indexes can be derived and monitored online to the system operator. Different (faster than real-time) stability programs for a number of contingencies can then be run to evaluate risks and margins.

Different kinds of "intelligent" load shedding can be ordered more or less automatically from the SCADA/EMS system in case of energy shortage on the electricity market or other limitations in the power system operation, that can be planned in advance. With access to wide-area measurements, such a system can be made adaptive to cope with the actual system conditions, such as load flow pattern and voltage levels.

It is necessary to distinguish between protection and SCADA/EMS, since this reflects the organizational structure of utility and grid company companies: The responsibilities for protection and for SCADA/EMS are given to different departments within the company.

9.3.3 Architectures of Wide-Area Protection System

Since the requirements for a wide-area protection system can vary from one utility company to another, the architecture for such a system must be designed according to what technologies the utility possesses at the given time. Also, to avoid becoming obsolete, the design must be chosen to fit the technology migration path that the utility in question will take. Three major design approaches are discussed below.

1. **Enhancements to SCADA/EMS**

At one end of the spectrum, enhancements to the existing EMS/SCADA can be made. These enhancements are aimed at two key areas: information availability and information interpretation. Simply put, if the operator has all vital information at his fingertips and good analysis facilities, he can operate the grid in an efficient way. For example, with better analysis tool for voltage instability, the operator can accurately track the power margin across an interface, and thus can confidently push the limit of transfer across that interface.

SCADA/EMS system capability has been greatly improved during recent years, due to improved communication facilities and highly extended data handling capability. New transducers such as the PMUs can provide time-synchronized measurements from all over the grid. Based on these measurements, improved state estimators can be derived.

Advanced algorithms and calculation programs that assist the operator can also be included in the SCADA system, such as "faster than real-time simulations" to calculate power transfer margins based on contingencies. The possibilities of extending the SCADA/EMS system with new functions tend to be limited. Therefore, it might be relevant to provide new SCADA/EMS functions as "stand alone" solutions, more or less independent of the ordinary SCADA/EMS system. Such functions could be load shedding, due to lack of generation or due to excessive market price.

2. "Flat Architecture" with System Protection Terminals

Protection devices or terminals are traditionally used in protecting equipment (lines, transformers, etc.). Modern protection devices have sufficient computing and communications capabilities that they are capable of performing beyond the traditional functions. When connected together via communications links, these devices can process intelligent algorithms (or "agents") based on data collected locally or shared with other devices.

Powerful, reliable, sensitive, and robust, wide-area protection systems can be designed based on decentralized, especially developed interconnected system protection terminals.

These terminals are installed in substations, where actions are to be made or measurements are to be taken. Actions are preferably local, i.e., transfer trips should be avoided, to increase security. Relevant power system variable data is transferred through the communication system that ties the terminals together. Different schemes, e.g., against voltage instability and against frequency instability, can be implemented in the same hardware.

Different layers of protection can be used, compared with the different zones of a distance protection. The voltage is for example measured in eight 400 kV nodes in a protection system against voltage instability. In a certain node, a certain action is taken if six of the eight voltages are low (e.g., <380 kV); or four of the eight voltages are very low (e.g., <370 kV); or the local voltage is extremely low (e.g., <360 kV).

Using the communication system, between the terminals, a very sensitive system can be designed. If the communication is partially or totally lost, actions can still be taken based on local criteria. Different load shedding steps that take the power system response into account, in order not to overshed, can easily be designed.

Protection systems against voltage instability can use simple binary signals such as "low voltage" or more advanced indicators such as power transfer margins based on the VIP algorithm or modal analysis.

The solution with interconnected system protection terminals for future transmission system applications is illustrated in Fig. 9.6 for protection against voltage instability; similar illustration can be done for angular instability.

3. Multilayered Architecture

While the above two designs attempt to extend the "reach" of existing control domains (protection terminal being one domain and EMS being the other), there is no guarantee that the end solution will be comprehensive. A comprehensive solution is one that integrates the two control domains, protection devices and EMS. Such a solution is depicted in Fig. 9.7.

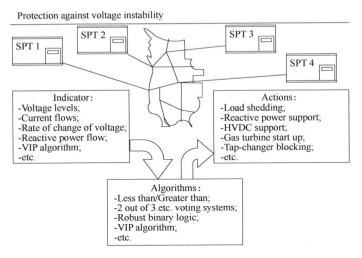

Fig. 9.6 Terminal-based wide-area protection system against voltage instability

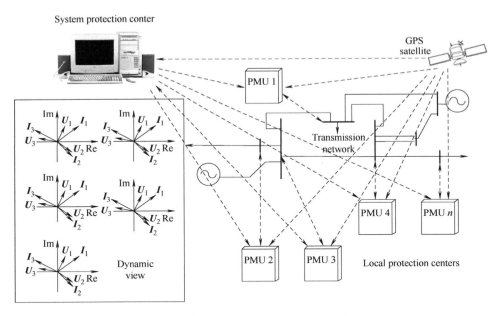

Fig. 9.7 Multilayered wide-area protection architecture

There are up to three layers in this architecture. The bottom layer is made up of PMUs, or PMUs with additional protection functionality. The next layer up consists of several local protection centers (LPCs), each of which interfaces directly with a number of PMUs. The top layer, the system protection center (SPC), acts as the coordinator for the LPCs.

Designing the three-layered architecture can take place in several steps. The first step should aim at achieving the monitoring capability, e.g. a WAMS. WAMS is the most common application based on PMUs. These systems are most frequent in North America, but are emerging all around the world.

The main purpose is to improve state estimation, postfault analysis, and operator information. In WAMS applications, a number of PMUs are connected to a data concentrator, which basically is a mass storage, accessible from the control center, according to Fig. 9.8.

Starting from a WAMS design, a data concentrator can be turned into a hub-based local protection center (LPC) by implementing control and protection functions in the data concentrator (Fig. 9.9).

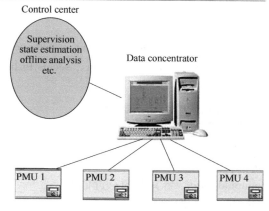

Fig. 9.8 WAMS design

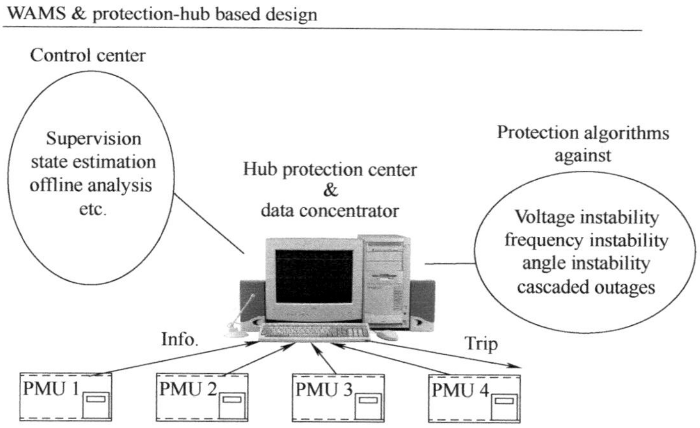

Fig. 9.9 Hub-based wide-area protection design

A number of such local protection centers can then be integrated into a larger system wide solution with a SPC at the top; see Fig. 9.7. With this solution the local protection center forms a system protection scheme (SPS), while the interconnected coordinated system forms a defense plan.

9.3.4 Discussion

The meaning of wide-area protection, emergency control, and power system optimization may vary dependent on people, utility, and part of the world, although the basic phenomena to be resolved are the same. Therefore, standardized and accepted terminology is important.

The solution to counteract the same physical phenomenon might vary extensively for different applications and utility conditions. A certain utility might wish to introduce a complete system to take care of a large number of applications, while others want to start with small installations of new technology in parallel with present systems. Some utilities want to do large amount of the studies, design, and engineering themselves, while others want to buy complete turnkey systems. It is important for any vendor in this area to supply solutions that fit with different utility organizations and traditions.

The potential to improve power system performance using smart control, as a complement to high-voltage equipment installations, seems to be great for many power grids.

VOCABULARY

1. phaser n. 移相器；相位器
2. merit n. 优点，价值
3. nominal adj. 名义上的；近似的
4. aliasing n. 混淆现象；折叠失真
5. recursive adj. 递归的；循环
6. leased adj. 租用的
7. coarse adj. 粗略的；粗糙的；粗俗的；下等的
8. navigational adj. 导航；航行的，航运的
9. furnish vt. 提供；供应；装备
10. antenna n. 天线；触角
11. algorithm n. 算法，运算法则
12. rotate vi. 旋转；循环
13. velocity n. 速率；迅速；周转率
14. prevailing adj. 一般的，最普通的；盛行很广的；占优势的；流行的
15. iterative adj. 迭代的；重复的，反复的
16. skew n. 斜交；歪斜 adj. 斜交的；歪斜的
17. validate vt. 确认；使生效；证实
18. matrix n. 矩阵；模型
19. dimensionality n. 维度；幅员
20. trajectory n. 轨道，轨线；弹道
21. derivatives n. 派生物；引出物；派生词
22. sabotage n. 蓄意破坏；破坏活动
23. shed vt. 流出；摆脱；散发
24. spectrum n. 光谱；频谱
25. fingertip n. 指尖；指套
26. robust adj. 强健的；健康的；粗野的
27. turnkey n. 成套系统

NOTES

1. For example, when the power system is undergoing electromechanical oscillations during power swings, the wave-forms of voltages and currents are not in steady state, and neither is the frequency of the power system at its nominal value.

例如，当电力系统在功率波动时发生机电振荡，电压和电流的波形不是稳定状态，并且电力系统的频率也不是额定值。

2. It should be noted that the waveform is sampled by the measurement system continuously, and each time a new sample is acquired, a new phasor is obtained with a data window including the latest sample.

应该指出，由测量系统连续采样得到波形，每当得到一个新的采样值，就可以从包含这个最新采样值的数据窗得到一个新的相位。

3. When several voltages and currents in a power system are measured and converted to phasors in this fashion, they are on a common reference if they are sampled at precisely the same instant.

当以这种方式对电力系统中的电压和电流进行测量及相位转换时，如果它们正好在同一时刻采样，那么它们就具有共同的参考相量。

4. Fiberoptic links, where available, could be used to provide high precision synchronization signals, if a dedicated fiber is available for this purpose.

可以的话，光缆线路能用来提供高精度的同步信号。为此可提供专用光纤。

5. The microprocessor determines the positive sequence phasors according to the recursive algorithm described previously, and the timing message from the GPS, along with the sample number at the beginning of a window, is assigned to the phasor as its identifying tag.

根据前面所述的递归算法微处理器决定正序相位，并把 GPS 发出的时间信息和从一个时间窗口开始的采样次数赋给相位作为识别标记。

6. The state (vector) is a collection of all the positive sequence voltage phasors of the network, and, from the time the first measurement is taken to the time when the state estimate is available, several seconds or minutes may have elapsed.

状态（矢量）是一个网络的所有正序电压相量的集合，从第一次测量到能进行状态估计时，几秒钟或几分钟可能已经过去了。

7. One of the most obvious applications of the instability prediction technique is to provide an improved adaptive out-of-step protection function, which adapts to changing power system conditions.

失稳预测技术最显著的应用之一是提供一种改进的自适应失步保护功能，这与不断变化的电力系统状况相适应。

8. Maximum benefit from available hardware that have passed different kinds of EMC and environmental tests, as well as software functions available from other protection terminals, has to be made.

（我们）可以从通过了各种电磁兼容和环境测试的硬件及安装在其他保护终端上的软件来获得最大的效益。

9. If the communication is partially or totally lost, actions can still be taken based on local criteria. Different load shedding steps that take the power system response into account, in order not to overshed, can easily be designed.

如果通信部分或完全丧失，仍可根据局部标准进行操作。为了不至于过度切负荷，考虑到电力系统的响应可以容易地设计不同的甩负荷步骤。

10. The meaning of wide-area protection, emergency control, and power system optimization may vary dependent on people, utility, and part of the world, although the basic phenomena to be resolved are the same.

尽管要解决的基本现象是相同的，但广域保护、紧急控制和电力系统优化的含义可能会因人、设施功能、地区等的不同而有所不同。

Chapter 10
Electronic Devices Simulation

10.1 Introduction

Electrical power systems are combinations of electrical circuits and electromechanical devices like motors and generators. Engineers working in this discipline are constantly improving the performance of the systems. Requirements for drastically increased efficiency have forced power system designers to use power electronic devices and sophisticated control system concepts that tax traditional analysis tools and techniques. Further complicating the analyst's role is the fact that the system is often so nonlinear that the only way to understand it is through simulation. Land-based power generation from hydroelectric, steam, or other devices is not the only use of power systems. A common attribute of these systems is their use of power electronics and control systems to achieve their performance objectives.

SimPowerSystems software is a modern design tool that allows scientists and engineers to rapidly and easily build models that simulate power systems. It uses the Simulink environment, allowing you to build a model. Not only can you draw the circuit topology rapidly, but your analysis of the circuit can include its interactions with mechanical, thermal, control, and other disciplines. This is possible because all the electrical parts of the simulation interact with the extensive Simulink modeling library. Since Simulink uses the MATLAB computational engine, designers can also use MATLAB toolboxes and Simulink blocksets. SimPowerSystems software belongs to the Physical Modeling product family and uses similar block and connection line interface.

The following statements do not represent an exhaustive list of pros about computer simulation, but they can certainly be thought as a "help list" available during the negotiations:

1) Here is an argument: simulation can avoid waste of time and money. With its inherent iterative power, SPICE covers numerous application cases in which you could easily detect any design flaw or product weakness. The stability of a closed-loop SMPS represents a typical application when some key feedback elements are moving (i. e., the variable load that affects a pole) or start to degrade with temperature and aging (as the electrolytic equivalent series resistor). Moreover, design ideas can also be tested or assessed in a snapshot through a computer and, if they are worth trying, further refined in the lab.

2) You can start to work on a project by downloading components models and become familiar with the key elements, before going to the bench or waiting for the samples to be delivered. Once

they arrive, you will have already gained insight and the debug phase on the bench will clearly have been beneficial!

3) Simulate test measurements whenever you do not own the adequate equipment. Bandwidth measurements represent a good example. If you cannot afford a network analyzer, then a proven small-signal model can start to help you refine your feedback loop. When run on the final prototype, stability assessments will be faster and more efficient.

4) Power libraries are safe: They let you experiment "what if" when amperes and kilovolts are flowing in the circuit without blowing up in the case of a wrong connection! Also, they let you see how your design reacts to a short-circuit of the optocoupler, or the opening of a resistor. SPICE can give you the answer.

10.2 Simulation Program with Integrated Circuit Emphasis (PSPICE)

10.2.1 Introduction of PSPICE Fundamentals

A schematic input file is created from circuit schematic. From the schematics file, a netlist is created automatically. Using PSPICE A/D, all the voltages in the circuit are obtained. The simulation results are processed using PROBE. Another way of performing PSPICE simulations is to start with a circuit file. In circuit files, the user assigns node numbers, specifies the elements connected to the various nodes, the type of simulation to be performed, and the output to be printed or plotted.

For many circuit simulations, working from the circuit schematic is much more convenient than performing the simulations using circuit files. However, for circuits that are very large and require advanced commands, the generation of circuit files is often necessary. This sector describes PSPICE simulations using circuit files.

A general SPICE circuit file program consists of the following components: title, element statements, control statements, and end statements.

The following two sections will discuss the element and control statements.

10.2.2 Element Statements

The element statements specify the elements in the circuit. The element statement contains the element name, the circuit nodes to which each element is connected, and the values of the parameters that electrically characterize the element.

The element name must begin with a letter of the alphabet that is unique to a circuit element, source, or subcircuit. Tab. 10.1 shows the beginning alphabet of an element name and the corresponding element.

Circuit nodes are positive integers. The nodes in the circuit need not be numbered sequentially. Node 0 is predefined for ground of a circuit. To prevent error messages, all nodes must be connected to at least two elements. Element values can be an integer, floating pointer number, integer floating point followed by an exponent, or floating point or integer followed by scaling factors. Any character after the scaling factor abbreviation is ignored in SPICE. For example, a 5000 Ohm resistor can be written as 5000 or 5000.00 Ohm, 5 K, 5 E3, 5 KOhm, or 5 KR.

Tab. 10.1 Element name and corresponding element

First letter of element name	Circuit element, sources, and subcircuit
B	GaAs MES field-effect transistor
C	Capacitor
D	Diode
E	Voltage-controlled voltage source
F	Current-controlled current source
G	Voltage-controlled current source
H	Current-controlled voltage source
I	Independent current source
J	Junction field-effect transistor
K	Mutual inductors (transformers)
L	Inductor
M	MOS field-effect transistor
Q	Bipolar junction transfer
R	Resistor
S	Voltage-controlled switch
T	Transmission line
V	Independent voltage source
X	Subcircuit

10.2.3 Design Manager

1. Introduction of Design Manager

Design Manager lets you manage design files and their dependencies as one unit. It has powerful file management features that let you perform file management activities on entire designs, as well as archive and restore entire designs for portability and storage. You can easily browse your design for file relationships within a dedicated workspace. Design Manager works very closely with the Design Journal feature in Schematics. Combining Design Manager file management and archiving functions with Design Journal design analysis and historical record keeping functions, provides maximum capability for design management, documentation, and tracking.

The Design Manager is automatically opened and minimized when you open any MicroSim program. You can also open it to view and manage files when other MicroSim programs are not open.

To activate Design Manager from within a MicroSim program. On the task bar, click the MicroSim Design Manager icon (Fig. 10.1).

Chapter 10 Electronic Devices Simulation

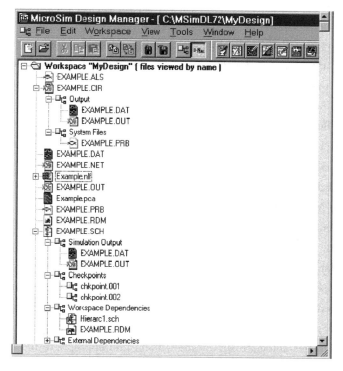

Fig. 10.1 MicroSim Design Manager

To activate Design Manager when no MicroSim programs are open:

1) On the task bar, click the Start button;

2) Point to Programs;

3) Point to the MicroSim entry;

4) Click MicroSim Design Manager.

The Category view provides a quick way to browse your design and easily identify design, derived, and other associated files.

When you select the Category view, Design Manager displays the files in categories, according to file type. For instance, all Schematic. sch files are listed in the Schematics category, while all simulation . dat and . out files are listed in the Simulation Output category.

When you copy or move files into a workspace, Design Manager automatically displays files in their proper category.

Design Manager also uses sub-categories to show the relationship between a design file, its dependencies, derived files, and other associated files.

For instance, if a Schematics top-level . sch file has dependencies (such as hierarchical sub-schematics), those dependencies are listed in either the Workspace Dependencies or External Dependencies sub-category, under the . sch top-level file name. External dependencies show a reference to their location.

If checkpoint schematics have been created for the top-level file, they are listed in the

Checkpoints sub-category.

2. Opening Applications from Design Manager

You can open any installed application from Design Manager. There are two ways to open a MicroSim application from Design Manager. To open MicroSim applications from the Tools menu, select the application that you want to open. In the workspace, double-clicking the icon for a file created by the MicroSim application that you want to open.

(1) Opening Files from Design Manager

You can open files from Design Manger, in their proper application. There are two ways to do so: using the Open command in the File menu; double-clicking the file icon in the workspace; to open a file using the File menu.

(2) About File Management

One of Design Manager great strengths is its flexible file management abilities. You can manage design files as one unit using Workspace commands, or you can use the Edit commands to perform file management similar to Windows Explorer.

Workspace and external dependencies can be included in all file management operations, depending on the dependency level. You can also include output and checkpoint files.

When files are copied, moved, or saved into a workspace, using either Workspace or Edit commands, Design Manager automatically arranges them into appropriate file-type categories.

Windows Explorer (when opened or refreshed) reflects any Design Manager file management activity.

To get optimum use of Design Manager: Create a workspace (folder) for each design.

Because Design Manager can perform operations on a top-level file and all of its dependencies simultaneously, perform file management functions within Design Manger, rather than Windows Explorer.

Design Manager tracks top-level files and their dependency relationships. If you cut, move, or delete dependencies using Windows Explorer, Design Manager will not be notified of the activity and will lose track of the relationships.

1) Select the file in Design Manager that you want to open.

2) From the File menu, select Open.

To open non-MicroSim applications, in the Workspace, double click the icon for a file created by the application that you want to open.

(3) Operations of the Workspace

To create a new workspace from Schematics:

1) From the File menu, select Save As.

2) In the Save In box, enter a file path that does not exist.

3) In the Name box enter a file name.

4) Click Save.

To create a new workspace from Design Manager:

1) From the File menu, select New Workspace.

2) In the Name box, type the workspace name.

3) In the Location box, enter a file path that does not exist.

4) Optionally, in the Description box, enter descriptive text.

5) Click the Create button.

(4) Copying All Files in the Workspace

When copying all files in the workspace, all dependency levels are available. If workspace or external dependencies have external dependencies, those external dependencies will be copied too.

To copy all files in the workspace:

1) From the Workspace menu, select Copy.

2) In the Destination box, enter the workspace path name that is to receive the copied files.

3) In the Copy frame, select the All Workspace Files button.

4) In the Dependencies box, choose the dependency level to be copied.

5) If you want to include checkpoint files, select the Schematics Checkpoints checkbox. Checkpoint files are created when using the Design Journal feature.

6) Click the Copy button.

You can also copy top-level files and their dependencies by pressing C and dragging the selected top-level files to a destination workspace. You are prompted for the dependency level when you drop the selected files.

You can have multiple workspaces, in their own windows, open simultaneously and perform file management activities among them.

(5) To Open an Existing Workspace

1) From the File menu, select Open Workspace;

2) Enter the file name and path;

3) Click Open.

To close a workspace, from the File menu, select Close Workspace.

Following is a simplified description of a typical Design Journal process:

1) A working schematic is created, developed to an acceptable reference level, and simulated.

2) The first checkpoint is created from the working schematic. The first checkpoint schematic is a copy of the current state of the working schematic. As long as the first checkpoint schematic is not changed, it preserves the integrity of the first reference point.

3) After the checkpoint is created, a description is entered into the checkpoint schematic to document pertinent facts about the checkpoint. The description is saved as part of the checkpoint schematic and cannot be edited separately from the schematic. Include a purpose statement for the checkpoint as part of the description.

4) What-if analysis or other modifications are performed on the working schematic.

5) The modified working schematic is simulated.

6) The traces of the working schematic and the first checkpoint schematic are compared using probe.

7) Working schematic is developed to a second level of acceptable reference.

8) A second checkpoint is created and documented. The second checkpoint schematic is a copy of the working schematic at its second reference level.

9) The iterative process of modifying the working schematic, simulating, and comparing it against any or all checkpoints simultaneously, and documenting each checkpoint, continues until the design is finished.

The collection of all checkpoints and the final working schematic form the Design Journal. The Design Journal is a chronological history of the design.

10.2.4 Device Model

The .MODEL statement specifies a set of device parameters that can be referenced by elements or devices in a circuit. The general form of the .MODEL statement is:

```
MODE L MODE L_NAME MODE L TYPE PARA METER_NAME = VALUE
```

Where

MODEL_NAME is a name for which devices use to reference a particular model. The model name must start with a letter. To avoid confusion, it is advisable to make the first character of the model_name identical with the first character of the device name.

MODEL_TYPE refers to the device type, which can be active or passive. The reference model may be available in the main circuit file, or accessed through a .INC statement, or may be in a library file. A device cannot reference a model statement that does not correspond to that type of model. It is possible to have more than one model of the same type in the circuit file, but they must have different model names; and PARAMETER_VALUES follow the model type. The model parameter values are enclosed in parenthesis. It is not required to list all the parameters values of the device. Parameters not specified are assigned default values.

In general, the .MODEL statement should adhere to the following rules:

1) More than one .MODEL statement can appear in a circuit file and each .MODEL statement should have a different model name. For example, the following models of a MOSFET are valid.

```
M1 1 2 0 0 MOD1 L=5U W=10U
M2 1 2 0 0 MOD2 L=5U W=15U
.MODEL MOD1 NMOS (VTO=1.0 KP=200)
.MODEL MOD2 NMOS (VTO=1.0 KP=150)
```

There are two model names, MOD1, MOD2 for model type NMOS.

2) More than one device of the same type may reference a given model using the .MODEL statement. For example:

```
D1 1 2 DMOD
D2 2 3 DMOD
.MODEL DMOD D (IS=1.0E-14 CJP=0.3P VJ=0.5)
```

Two diodes, D1 and D2, reference the given diode model DMOD.

3) A device cannot reference a model statement that does not correspond to the device. For example, the following statements are incorrect.

```
R1 1 2 DMOD
.MODEL DMOD D (IS =1.0E-12)
Resistor R1 cannot reference a diode model DMOD.
Q1 3 2 1 MMDEL
.MODEL MMDEL NMOS (VTO =1.2)
```

The bipolar transistor cannot reference the NMOS transistor.

4) A device cannot reference more than one model in a netlist.

In the following sections, we shall discuss the .MODEL statements for both passive (R, L, C) and active elements (D, M, Q).

1. Resistor Models

The basic description of passive elements was expressed in terms of element name, nodal connections, and component value. To model a resistor, two statements are required. The general format is:

```
Rname NODE1 NODE2 MODEL_NAME? R_VALUE
.MODEL MODEL_NAME? RES [MODEL_PARAMETER]
```

Where

MODEL_NAME is a name preferably starting with the character R. It can be up to eight characters long.

RES is the specification for PSPICE model type associated with resistors.

MODEL_PARAMETERS are parameters that can vary.

PSPICE uses the model parameters to calculate the resistance using the following equations:

$$\text{R\_model}(T) = \text{R\_value} * R[1 + \text{TC1}(T - T_{nom}) + \text{TC2}(T - T_{nom})^2] \quad (10\text{-}1)$$

or

$$\text{R\_model} = \text{R\_value} * R[1.01]^{\text{TCE}(T - T_{nom})} \quad (10\text{-}2)$$

Where T is the temperature at which the resistance needs to be calculated.

Tab. 10.2 Resistor model parameter description their default value

Model Parameter	Description	Default value	Unit
R	Resistance multiplier	1	
TC1	Linear temperature coefficient	0	℃$^{-1}$
TC2	Quadratic temperature coefficient	0	℃$^{-2}$
TCE	Exponential temperature coefficient	Default value	%℃

Equation (10-1) uses the linear and quadratic temperature coefficients of the resistor. The coefficients TC1 and TC2 are specified in the .MODEL statement. Equation (10-2) is used if the resistance is exponentially dependent on temperature.

For example, the statements:

```
R1 4 3 RMOD1 5K
.MODEL RMOD1 RES(R=1,TC1=0.00010)
```

describe a resistor with model name RMOD1. The .MODEL statement specifies that the resistor R1 has a linear temperature coefficient of +100 ppm/℃.

The statement:

```
R2 5 4 RMOD2 10K
.MODE L R MOD2 RES (R=2,TCE=0.0010)
```

describes a resistor whose value with respect to temperature is given by the expression:
$$R2(T) = 10000 * 2 * (1.01)^{0.001(T-T_{nom})} \quad (10\text{-}3)$$
Where T_{nom} is the normal temperature that can be set with T_{nom} option.

2. Capacitor Models

Whereas the resistors are modeled to be temperature dependent, capacitors can be both temperature and voltage dependent. In addition, capacitors can have initial voltage impressed on them. The element description and the model statements have the feature to incorporate the above influences on capacitors. The general format for modeling capacitors is:

```
CNAME NODE+NODE-MODEL_NAME VALUE IC=INITIAL_VALUE.
MODEL MODEL_NAME CAP MODEL_PARAMETERS
```

Where

MODEL_NAME is a name preferably starting with character C. It can be up to eight characters long;

CAP is the PSPICE specification for the model type associated with capacitors;

MODEL_PARA METERS are parameters that can be used to describe the capacitance value with respect to changes in temperature and voltage.

PSPICE uses the model parameters to calculate the capacitance at a particular temperature T, and voltage V, using the following expression:
$$C(V,T) = C\_value * C[1 + VC1*V + VC2*V^2] * [1 + TC1(T-T_{nom}) + TC2(T-T_{nom})^2]$$
Where T_{nom} is nominal temperature set by T_{nom} option.

For example, the statements:

```
CBIAS 5 0 CMODEL 20e-6 IC=3.0
.MODEL CMODEL CAP (C=1,VC1=0.0001,VC2=0.00001,TC1=-0.000006)
```

describe a capacitance that is a function of both voltage V and temperature T and whose value is given as

$$C(V,T) = 20.0 * 10^{-6}[1 + 0.001V + 0.00001V^2] * [1 - 0.000006(T - T_{nom})] \quad (10\text{-}4)$$

3. Inductor Models

Inductors are current and temperature independent. They are thus modeled similarly to capacitors.

4. Diode Models

Diode models take into account the forward and reverse bias characteristics of real diodes, the junction capacitance, the ohmic resistance of the diode, the temperature effects on the diode characteristics, and leakage current and high level injection of real diodes.

5. Bipolar Junction Transistor Models

The PSPICE model of bipolar junction transistors is based on the integral charge-control of Gummel and Poon. For large signal analysis, the Ebers and Moll transistor model can be used. The general format for modeling bipolar transistors is:

```
QNAME NC NB NE NS MODE L_NAME < AREA_VALUE >.
MODEL MODEL_NAME TRANSISTOR_TYPE
MODEL_PARAMETERS
```

where

NC, NB, NE, and NS are the node numbers for the collector, base, emitter, and substrate, respectively;

MODEL_NAME is a name preferably starting with the character Q. It can also be up to eight characters long;

AREA VALUE is the size of the transistor. It represents the number of transistors paralleled together. The parameters that are affected by the area factor are: IS, IKR, RB, RE, RC, and CJE;

TRANSISTOR_TYPE can be either NPN or PNP;

MODEL_PARAMETERS are the parameters that model the bipolar junction transistor characteristics.

6. MOSFET Models

Depending on the level of complexity, MOSFETS can be modeled by Shichman-Hodges model, geometry-based analytic model, semi-empirical short-channel model, or Berkeley short-channel IGFET model (BSIM). The general format for modeling MOSFETS is:

```
MNAME ND NG NS NB MODEL_NAME DEVICE_PARAMETERS.
MODEL MODEL_NAME TRANSISTOR_TYPE MODEL_PARAMETERS
```

where

ND, NG, NS, and NB are the node numbers for the drain, gate, source, and substrate, respectively;

MODEL_NAME is a name preferably starting with the character M. It can be up to eight characters long;

DEVICE_PARAMETERS are optional parameters that can be provided for L (length), W (width), AD (drain diffusion area), AS (source diffusion area), PD (perimeter of drain diffusion), and RS (perimeter of source diffusion). NRD, NRS, NRG, and NRB are the relative resistivities of the drain, source, gate, and substrate in squares. M is the device "multiplier." Its default value is one. It simulates the effect of equivalent MOSFETS connected in parallel;

TRANSISTOR_TYPE can either be NMOS or PMOS;

MODEL_PARAMETERS selected depends on the MOSFET model that is being used. The LEVEL parameter is used to select the appropriate model. The following models are available:

LEVEL = 1 is for Shichman-Hodges model;

LEVEL = 2 is a geometry-based analytic model.

10.2.5 Library File

Models and subcircuits of devices and components exist in PSPICE. There are more than 5000 device models available that PSPICE users can use for simulation and design. The models exist in different libraries of the PSPICE package. The reader should consult PSPICE manuals for models available for various electronic components.

The LIB statement is used to reference a model or a subcircuit that exists in another file as a library. The general format for the LIB command is:

    .LIB FILENAME.LIB

Where

FILENAME. LIB is the name of the library file.

If the FILENAME. LIB is left off then the default file is NOM. LIB. The latter library, depending on the version of PSPICE you are running, will contain devices or names of individual libraries.

One can also set up one's own library file using the file extension .LIB. The device models and subcircuits can be placed in the file. One can access the individually created libraries in the same way as the PSPICE supplied libraries are accessed, i.e., by using:

    .LIB FILENAME.LIB.

One should be careful not to give the individually created library file the same name as the ones supplied by PSPICE. The following example explores the use of models in a diode circuit.

10.3 MATLAB/Simulink

10.3.1 Introduction of MATLAB

MATLAB is a numeric computation software for engineering and scientific calculations. The name MATLAB stands for matrix laboratory.

MATLAB is primarily a tool for matrix computations. It was developed by John Little and Cleve

Moler of MathWorks, Inc. MATLAB was originally written to provide easy access to the matrix computation software packages LINPACK and EISPACK. MATLAB is a high-level language whose basic data type is a matrix that does not require dimensioning. There is no compilation and linking as is done in high-level languages, such as C or FORTRAN. Computer solutions in MATLAB seem to be much quicker than those of a high-level language such as C or FORTRAN. All computations are performed in complex-valued double precision arithmetic to guarantee high accuracy.

MATLAB has a rich set of plotting capabilities. The graphics are integrated in MATLAB. Since MATLAB is also a programming environment, a user can extend the functional capabilities of MATLAB by writing new modules.

MATLAB has a large collection of toolboxes in a variety of domains. Some examples of MATLAB toolboxes are control system, signal processing, neural network, image processing, and system identification. The toolboxes consist of functions that can be used to perform computations in a specific domain.

10.3.2 SimPowerSystems Block Libraries

1. Overview of SimPowerSystems Libraries

SimPowerSystems libraries contain models of typical power equipment such as transformers, lines, machines, and power electronics. These models are proven ones coming from textbooks, and their validity is based on the experience of the Power Systems Testing and Simulation Laboratory of Hydro-Québec, a large North American utility located in Canada. The capabilities of SimPowerSystems software for modeling a typical electrical system are illustrated in demonstration files. And for users who want to refresh their knowledge of power system theory, there are also self-learning case studies.

The SimPowerSystems main library, powerlib, organizes its blocks into libraries according to their behavior. To open this library, type powerlib in the MATLAB Command Window. The powerlib library window displays the block library icons and names. Double-click a library icon to open the library and access the blocks. The main powerlib library window also contains the Powergui block that opens a graphical user interface for the steady-state analysis of electrical circuits.

2. Nonlinear Simulink Blocks for SimPowerSystems Models

The nonlinear Simulink blocks of the powerlib library are stored in a special block library named powerlib_models. These masked Simulink models are used by SimPowerSystems software to build the equivalent Simulink model of your circuit. You can also access SimPowerSystems libraries through the Simulink Library Browser. To display the Library Browser, click the Library Browser button in the toolbar of the MATLAB desktop or Simulink model window. Alternatively, you can type Simulink in the MATLAB Command Window. Then expand the Simscape entry in the contents tree.

3. Building the Electrical Circuit with powerlib Library

The graphical user interface makes use of the Simulink functionality to interconnect various electrical components. The electrical components are grouped in a library called powerlib.

1) Open the SimPowerSystems main library by entering the following command at the MATLAB prompt.

This command displays a Simulink window showing icons of different block libraries (Fig. 10.2).

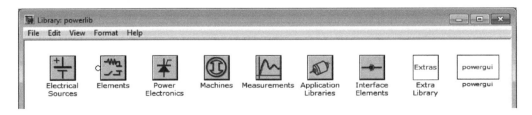

Fig. 10.2　Different block libraries

You can open these libraries to produce the windows containing the blocks to be copied into your circuit. Each component is represented by a special icon having one or several inputs and outputs corresponding to the different terminals of the component:

2) From the File menu of the powerlib window, open a new window to contain your first circuit and save it as circuit1.

3) Open the Electrical Sources library and copy the AC Voltage Source block into the circuit1 window.

4) Open the AC Voltage Source dialog box by double-clicking the icon and enter the Amplitude, Phase, and Frequency parameters according to the values shown in circuit to be modeled (Fig. 10.3). Note that the amplitude to be specified for a sinusoidal source is its peak value.

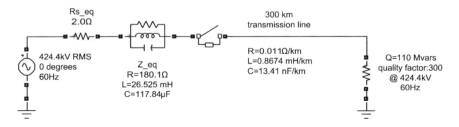

Fig. 10.3　Circuit to be modeled

5) Change the name of this block from AC voltage source to V_s (Fig. 10.4).

6) Copy the Parallel RLC Branch block, which can be found in the Elements library of powerlib, set its parameters as shown in Fig. 10.3, and name it Z_eq.

7) The resistance Rs_eq of the circuit can be obtained from the Parallel RLC Branch block. Duplicate the Parallel RLC Branch block, which is already in your circuit1 window. Select R for the Branch Type parameter and set the R parameter. Once the dialog box is closed, notice that the L and

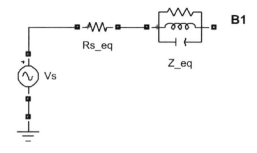

Fig. 10.4　Dragging lines

C components have disappeared so that the icon now shows a single resistor.

Note: With the Branch Type parameter set to RLC, setting L and C respectively to inf and zero in a parallel branch changes automatically the Branch Type to R and produces the same result. Similarly, with the Series RLC Branch block, setting R, L, and C respectively to zero, zero, and inf eliminates the corresponding element.

8) Name this block Rs_eq.

9) Resize the various components and interconnect blocks by dragging lines from outputs to inputs of appropriate blocks.

10) To complete the circuit of Fig. 10.3, you need to add a transmission line and a shunt reactor. You add the circuit breaker later.

The model of a line with uniformly distributed R, L, and C parameters normally consists of a delay equal to the wave propagation time along the line. This model cannot be simulated as a linear system because a delay corresponds to an infinite number of states. However, a good approximation of the line with a finite number of states can be obtained by cascading several PI circuits, each representing a small section of the line.

A PI section consists of a series RL branch and two shunt C branches. The model accuracy depends on the number of PI sections used for the model. Copy the PI Section Line block from the Elements library into the circuit1 window, set its parameters as shown in Fig. 10.3, and specify one line section.

11) The shunt reactor is modeled by a resistor in series with an inductor. You could use a Series RLC Branch block to model the shunt reactor, but then you would have to manually calculate and set the R and L values from the quality factor and reactive power specified in Fig. 10.3.

Therefore, you might find it more convenient to use a Series RLC Load block that allows you to specify the active and reactive powers absorbed by the shunt reactor.

Copy the Series RLC Load block, which can be found in the Elements library of powerlib. Name this block 110 Mvar. Set its parameters as follows as Tab. 10.3.

Tab. 10.3 Parameters of Series RLC Load bolck

Parameter	Value
V_n/V	424.4e3
f_n/Hz	60
P/W	110e6/300 (quality factor = 300)
Q_L/vars	110e6
Q_C/vars	0

Note that, as no reactive capacitive power is specified, the capacitor disappears on the block icon when the dialog box is closed. Interconnect the new blocks as shown in Fig. 10.5.

12) You need a Voltage Measurement block to measure the voltage at node B1. This block is found in the Measurements library of powerlib. Copy it and name it U_1. Connect its positive input to

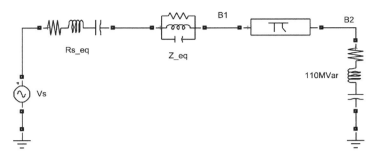

Fig. 10.5 Interconnect the new blocks

the node B1 and its negative input to a new Ground block.

13) To observe the voltage measured by the Voltage Measurement block named U_1, a display system is needed. This can be any device found in the Simulink Sinks library.

Open the Sinks library and copy the Scope block into your circuit1 window. If the scope were connected directly at the output of the voltage measurement, it would display the voltage in volts. However, electrical engineers in power systems use it to work with normalized quantities (per unit system). The voltage is normalized by dividing the value in volts by a base voltage corresponding to the peak value of the system nominal voltage.

14) Copy a Gain block from the Simulink library and set its gain as above.

Connect its output to the Scope block and connect the output of the Voltage Measurement block to the Gain block. Duplicate this voltage measurement system at the node B2, as shown as Fig. 10.6.

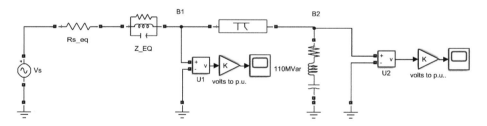

Fig. 10.6 Copy a Gain block

15) Add a Powergui block to your model. The purpose of this block is discussed in "Using the Powergui Block to Simulate SimPowerSystems Models".

16) From the Simulation menu, select Start.

17) Open the Scope blocks and observe the voltages at nodes B1 and B2.

18) While the simulation is running, open the V_s block dialog box and modify the amplitude. Observe the effect on the two scopes. You can also modify the frequency and the phase. You can zoom in on the waveforms in the scope windows by drawing a box around the region of interest with the left mouse button.

To simulate this circuit, the default integration algorithm (ode45) was used. However, for most

SimPowerSystems applications, your circuits contain switches and other nonlinear models. In such a case, you must specify a different integration algorithm.

4. Interfacing the Electrical Circuit with Other Simulink Blocks

The Voltage Measurement block acts as an interface between the SimPowerSystems blocks and the Simulink blocks. For the system shown above, you implemented such an interface from the electrical system to the Simulink system. The Voltage Measurement block converts the measured voltages into Simulink signals. Similarly, the Current Measurement block from the Measurements library of powerlib can be used to convert any measured current into a Simulink signal. You can also interface from Simulink blocks to the electrical system. For example, you can use the Controlled Voltage Source block to inject a voltage in an electrical circuit, as shown in the following Fig. 10. 7.

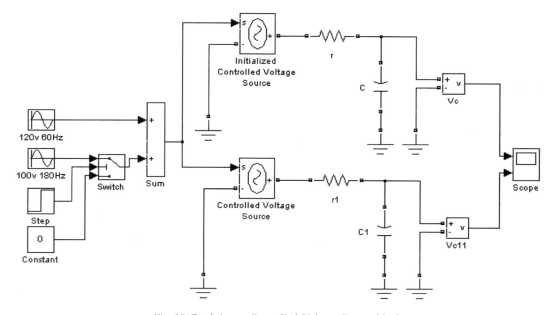

Fig. 10. 7　Inject a Controlled Voltage Source block

5. Electrical Terminal Ports and Connection Lines

SimPowerSystems modeling environment is similar to that of other products in the Physical Modeling family. Its blocks often feature both normal Simulink input and output ports and special electrical terminal ports:

1) Lines that connect normal Simulink ports are directional signal lines.

2) Lines that connect terminal ports are special electrical connection lines. These lines are nondirectional and can be branched, but you cannot connect them to Simulink ports or to normal Simulink signal lines.

3) You can connect Simulink ports only to other Simulink ports and electrical terminal portsor only to other electrical terminal ports.

4) Converting Simulink signals to electrical connections or vice versa requires using a SimPowerSystems block that features both Simulink ports and electrical terminal ports.

6. Block Orientation

The natural orientation of the blocks (that is, their orientation in the Element library) is right for horizontal blocks and down for vertical blocks. For single-phase transformers (linear or saturable), with the winding connectors appearing on the left and right sides, the winding voltages are the voltages of the top connector with respect to the bottom connector, irrespective of the block orientation (right or left). The winding currents are the currents entering the top connector. For three-phase transformers, the voltage polarities and positive current directions are indicated by the signal labels used in the Multimeter block. For example, U_{an_w2} means phase A-to-neutral voltage of the Y connected winding #2, I_{ab_w1} means winding current flowing from A to B in the delta-connected winding #1.

7. Basic Principles of Connecting Capacitors and Inductors

You have to pay particular attention when you connect capacitor elements together with voltage sources, or inductor elements in series with current sources. When you start the simulation, the software displays an error message if one of the following two connection errors are present in your diagram:

1) You have connected a voltage source in parallel with a capacitor, or a series of capacitor elements in series, like in the two examples below (Fig. 10.8).

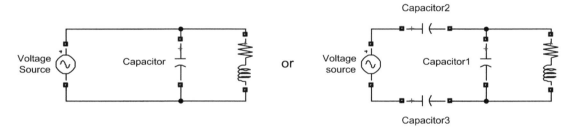

Fig. 10.8 Connected a voltage source

To fix this problem, you can add a small resistance in series between the voltage source and the capacitors.

2) You have connected a current source in series with an inductor, or a series of inductors connected in parallel, like in the example below.

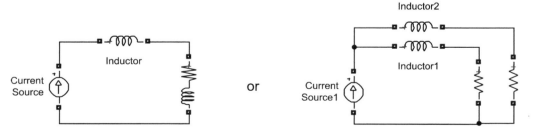

Fig. 10.9 Connected a current voltage source

To fix this problem, you can add a large resistance in parallel with the inductor.

8. Using the Powergui Block to Simulate SimPowerSystems Models

The Powergui block is necessary for simulation of any Simulink model containing SimPowerSystems blocks. It is used to store the equivalent Simulink circuit that represents the state-space equations of the SimPowerSystems blocks. You must follow these rules when using this block in a model:

1) Place the Powergui block at the top level of diagram for optimal performance. However, you can place it anywhere inside subsystems for your convenience; its functionality will not be affected.

2) You can have a maximum of one Powergui block per model.

3) You must name the block Powergui.

When you start the simulation, you will get an error if no Powergui block is found in your model. Position of the Powergui module is shown in the Fig. 10. 10.

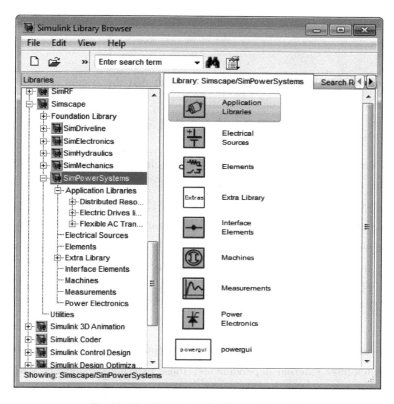

Fig. 10. 10   Position of the Powergui module

For more information on using the Library Browser, see "Library Browser" in the *Simulink Graphical User Interface* documentation.

VOCABULARY

1. negotiation   *n.* 谈判，磋商
2. electrolytic   *adj.* 电解的，由电解产生的
3. snapshot   *n.* 快照，急射，简单印象
4. probe   *n.* 探针，探测器
5. element   *n.* 成分，元件
6. methodology   *n.* 方法学，方法论
7. Design Manager   设计管理器
8. checkpoint   *n.* 检查点；检验点
9. double-clicking   双击
10. default value   默认值，缺省值

11. parenthesis　　n. 插入语，附带，插曲，圆括号
12. nodal　　adj. 节的，节点的，交点的
13. coefficient　　n. 系数
14. computation　　n. 计算，估计
15. toolbox　　n. 工具箱
16. nonlinear　　adj. 非线性的
17. toolbar　　n. 工具栏
18. content tree　　目录树
19. graphical　　adj. 绘成图画似的，绘画的
20. interface　　n. 分界面，接触面，界面
21. prompt　　vt. 提示，鼓动
22. dialog box　　对话框
23. wave propagation　　波形传播
24. voltage measurement　　电压测量
25. integration　　n. 积分
26. block　　n. 模块
27. gain　　n. 增益

NOTES

1. Further complicating the analyst's role is the fact that the system is often so nonlinear that the only way to understand it is through simulation.

事实上将来的分析任务会更加复杂，因为系统通常是强非线性的，了解系统的唯一方式是仿真。

2. For many circuit simulations, working from the circuit schematic is much more convenient than performing the simulations using circuit files.

对于很多电路仿真来说，用电路的原理图仿真要比用电路文件仿真方便得多。

3. Design Manager lets you manage design files and their dependencies as one unit. It has powerful file management features that let you perform file management activities on entire designs, as well as archive and restore entire designs for portability and storage.

设计管理器可以让你将设计文件及其从属项作为一个单元进行统一管理。它具有很强的文件管理功能，使你可以对整个设计进行文件管理，以及设计文档的归档与恢复，以便于携带和存储。

4. When copying all files in the workspace, all dependency levels are available. If workspace or external dependencies have external dependencies, those external dependencies will be copied too.

当在工作空间中复制所有的文件时，所有的从属文件也都是可复制的。当你选择了一个包含有外部从属文件的文件时，那些外部的从属文件也会被复制。

5. Diode models take into account the forward and reverse bias characteristics of real diodes, the junction capacitance, the ohmic resistance of the diode, the temperature effects on the diode characteristics, and leakage current and high level injection of real diodes.

二极管的模型考虑了实际二极管的正向和反向偏置特性、结电容、电阻欧姆值、温度对其特性的影响，以及实际二极管的漏电流和高电平注入等。

6. The capabilities of SimPowerSystems software for modeling a typical electrical system are illustrated in demonstration files. And for users who want to refresh their knowledge of power system theory, there are also self-learning case studies.

采用 SimPowerSystems 软件构建典型电气系统模型的性能在示例文件中进行了说明。对于那些想提高自身电力系统理论知识的用户来说，也有一些自学的例子。

7. The graphical user interface makes use of the Simulink functionality to interconnect various electrical components. The electrical components are grouped in a library called powerlib.

用户图形界面利用 Simulink 的功能将各种不同的电子元件相互连接在一起。这些电子元件组合在一起称为电力元件库。

8. This model cannot be simulated as a linear system because a delay corresponds to an infinite number of states.

这个模型不能用作线性系统仿真，因为一个延迟就对应无数个状态。

Chapter 11
Power System Simulation Softwares

11.1 Introduction

Software packages for power system analysis can be basically divided into two classes of tools: off-line simulation software and real-time simulation software. The off-line simulation softwares can be used to simulate the power system off line because it would take much time to achieve the simulation. On the contrary, the real-time simulation software can finish simulation in a short time and they can be used to simulate the power system on line.

There are many off-line power system analysis commercial softwares and educational/research-aimed softwares in the world. The off-line commercial softwares include Power System Simulation for Engineering (PSS/E), Digital Simulation and Electrical NeTwork calculation program (DIgSILENT, as written by PowerFactory), Power System Analysis and Synthesis Program (PSASP), Power System Calculation and Design (PSCAD), ElectroMagnetic Transient Program (EMTP), Electrical Transients including DC (EMTDC), Electrical Transient Analysis Program (ETAP) and ect. The off-line educational/research-aimed softwares include Power System Toolbox (PST), MatPower, Toolbox (VST), MatEMTP, SimPowerSystems (SPS), Power Analysis Toolbox (PAT), the Educational Simulation Tool (EST) and so on. Most of the educational/research-aimed softwares are based on MATLAB.

The most popular power system real-time simulation softwares is Real-Time Digital Simulation (RTDS). RTDS are used to replace the power system physical model dynamic simulation system more and more.

Three commercial power system analysis software (PSS/E, Power Factory and PSCAD), a educational-aimecl software (PSAT) and RTDS will be introduced in the following parts in this chapter.

11.2 Offline Simulation Softwares

11.2.1 Power System Simulation for Engineering (PSS/E)

1. Introduction to PSS/E

PSS/E is composed of a comprehensive set of programs for studies of power system transmission network and generation performance in both steady-state and dynamic conditions. Currently two primary simulations are used, one for steady-state analysis and one for dynamic simulations. PSS/E can be utilized to facilitate calculations for a variety of analyses, including:

1) Power flow and related network functions;
2) Optimal power flow;
3) Balanced and unbalanced faults;
4) Network equivalent construction;
5) Dynamic simulation.

PSS/E uses a graphical user interface that is comprised of all the functionality of state analysis; including load flow, fault analysis, optimal power flow, equivalency, and switching studies.

In addition, to the steady-state and dynamic analyses, PSS/E also provides the user with a wide rage of auxiliary programs for installation, data input, output, manipulation and preparation. Furthermore, one of the most basic premises of PSS/E is that the engineer can derive the greatest benefit from computational tools by retaining intimate control over their application.

2. Power Flow

A power flow study (also known as load-flow study) is an important tool involving numerical analysis applied to a power system. Unlike traditional circuit analysis, a power flow study usually uses simplified notation such as a one-line diagram and per-unit system, and focuses on various forms of AC power (i. e. reactive, real, and apparent).

Power flow studies are important because they allow for planning and future expansion of existing as well as non-existing power systems. A power flow study also can be used to determine the best and most effective design of power systems.

The PSS/E interface supports a variety of interactive facilities including:

1) Introduction, modification and deletion of network data using a spreadsheet;
2) Creation of networks and one-line diagrams;
3) Steady-state analyses (load flow, fault analysis, optimal power flow, etc.);
4) Presentation of steady-state analysis results.

3. Dynamics

The dynamic simulation program includes all the functionality for transient, dynamic and long term stability analysis. The dynamic simulation interface is operated as a separate program, currently independent of the PSS/E interface. This can be observed when going to a PSS/E program and viewing the dynamics as a separate program. The purpose of the dynamics is to facilitate operation of all dynamic stability analytical functions. The dynamics program, in addition to supporting the dynamics activities, also continues to support the traditional load flow interface through the LOFL activity.

11.2.2 PowerFactory (DIgSILENT)

1. Introduction

The calculation program PowerFactory, as written by DIgSILENT, is a computer aided engineering tool for the analysis of industrial, utility, and commercial electrical power systems. It has been designed as an advanced integrated and interactive software package dedicated to electrical power system and control analysis in order to achieve the main objectives of planning and operation optimization.

The name DIgSILENT stands for "DIgital SImuLation and Electrical NeTwork calculation program". DIgSILENT Version 7 was the world's first power system analysis software with an integrated graphical one-line interface. That interactive one-line diagram included drawing functions, editing capabilities and all relevant static and dynamic calculation features.

The PowerFactory package was designed and developed by qualified engineers and programmers with many years of experience in both electrical power system analysis and programming fields. The accuracy and validity of the results obtained with this package has been confirmed in a large number of implementations, by organizations involved in planning and operation of power systems.

In order to meet today's power system analysis requirements, the DIgSILENT power system calculation package was designed as an integrated engineering tool which provides a complete 'walk-around' technique through all available functions, rather than a collection of different software modules. The following key-features are provided within one single executable program:

1) PowerFactory core functions: definition, modification and organization of cases; core numerical routines; output and documentation functions.

2) Integrated interactive single line graphic and data case handling.

3) Power system element and base case database.

4) Integrated calculation functions (e.g. line and machine parameter calculation based on geometrical or nameplate information).

5) Power system network configuration with interactive or on-line access to the SCADA system.

6) Generic interface for computer-based mapping systems.

By using just a single database, containing all the required data for all equipment within a power system (e.g. line data, generator data, protection data, harmonic data, controller data), PowerFactory can easily execute any or all available functions, all within the same program environment. Some of these functions are load-flow, short-circuit calculation, harmonic analysis, protection coordination, stability calculation and modal analysis.

DIgSILENT PowerFactory has originally been designed as a complete package for the high-end user. Consequently, there are no special 'lightweight' versions, no cut-outs of a 'heavy' version. This does not, however, mean that non high-end users will find themselves at sea when using PowerFactory. The program is also friendly to the basic user. Users who are learning about power systems are able to easily and quickly perform load-flows and short-circuit calculations, without needing to immediately master the mathematical intricacies of the calculations. PowerFactory allows the user to learn primarily about power systems and not PC quirks—all that is required is a reasonable working knowledge of Windows applications such as Word and Excel.

The program is shipped with all of the engines and algorithms that are required for high-end use. The functionality that has been bought by a user is configured in a matrix, where the licensed calculation functions, together with the maximum number of busses, are listed as coordinates. In addition, there are options available which will allow the configuration and fine-tuning of the software according to the user's needs, for some of the functions.

In this manner, not every PowerFactory license contains all functionality described in this manual, but only those actually required, thereby reducing the complexity of the outset. As requirements dictate further functionality can be added to the license. The user thus does not have to learn a whole new interface for new functions, but merely uses new commands within the same environment. In addition, the original network data is used and only extra data, as may be required by the new calculation function, needs to be added.

2. General Concept

The general concept behind the program design and application can be described by means of the three basic integration characteristics that contribute to make PowerFactory a unique power system analysis tool:

(1) **Functional Integration**

DIgSILENT PowerFactory software is implemented as a single executable program, and is fully compatible with Windows 95/98/NT/2000/XP/Vista. The programming method employed allows for a fast walk around the execution environment, and eliminates the need to reload modules and update or transfer results between different program applications. As an example, the power flow, fault analysis, and harmonic load flow analysis tools can be executed sequentially without resetting the program, enabling additional software modules and engines or reading external data files.

(2) **Vertical Integration**

A special feature of the DIgSILENT PowerFactory software is the unique vertically integrated model concept. This allows models to be shared for all analysis functions and more importantly, for categories of analysis, such as Generation Transmission Distribution and Industrial. No longer are separate software engines required to analyze separate aspects of the power system, as DIgSILENT PowerFactory can accommodate everything within one integrated frame and one integrated database.

(3) **Database Integration**

DIgSILENT PowerFactory provides optimal organization of data and definitions required to perform any type of calculation, memorization of settings or software operation options. There is no need in tedious organization of several files for defining the various analysis aspects. The PowerFactory database environment fully integrates all data required for defining cases, operation scenarios, single-line graphics, outputs, run conditions, calculation options, graphics, user-defined models, etc. There is no need to keep and organize hundreds of files on hard disc, every thing you require to model and simulate a power system is integrated in a single database!

1) Single Database Concepts: All data for standard and advanced functions are organized in a single, integrated database. This is applied also for graphics, study case definitions, outputs, run conditions, calculation options, fault sequences, monitoring messages as well as user-defined models.

2) Project Management: All the data that defines a power system model and allows its

calculation is stored in so called Project folders within the database. Inside a Project, folders called Study Cases are used to define different studies of the system considering the complete network, only parts of it or variations on its current state. This project and study case approach to define and manage power system studies is a unique application of the object-oriented software principle. Standard software packages often require the user to create a large number of similar saved cases, with multiple nested directories for large complex networks and studies. However, DIgSILENT PowerFactory has taken a totally new approach, and introduced a structure that is both easy to use while avoiding redundancy.

3) Multi-User Operation: Multiple users each holding its own projects or working with data shared from other users are supported by a 'Multi-user' database operation. In this case the definition of access rights, user accounting and groups for data sharing are managed by a database administrator.

3. PowerFactory Simulation Functions

PowerFactory incorporates an impressive and continuously growing list of simulation functions including:

1) Load Flow and Fault Analysis, allowing meshed and mixed 1-, 2-, and 3-phase AC and/or DC networks;
2) Low Voltage Network Analysis;
3) Distribution Network Optimization;
4) IEC Cable Sizing;
5) Dynamic Simulation;
6) EMT Simulation;
7) Eigenvalue Analysis;
8) System Identification;
9) Protection Analysis;
10) Harmonic Analysis;
11) Reliability Analysis;
12) Voltage Stability Analysis;
13) Contingency Analysis;
14) Power Electronic Device Modeling;
15) Grounding;
16) A-D Interfacing;
17) Interface for SCADA/GIS/NIS;
18) Compatibility with other software systems such as PSS/E & PSS/U;
19) Multi-User Database and User Accounting;
20) Optimal Power Flow.

4. General Design of PowerFactory

In order to better understand how to use a program, it is useful to first get an idea of what the designers had in mind when they designed the user interface. In the next few paragraphs, we will attempt to explain what this philosophy is.

PowerFactory is intended to be initially used and operated in a graphical environment. That is, data entry is accomplished by drawing the network under study and then by editing the objects on the drawing canvas to assign data to them.

Fig. 11.1 shows how PowerFactory looks like when a project is active. It shows the Graphic window (up) and the Output window (below).

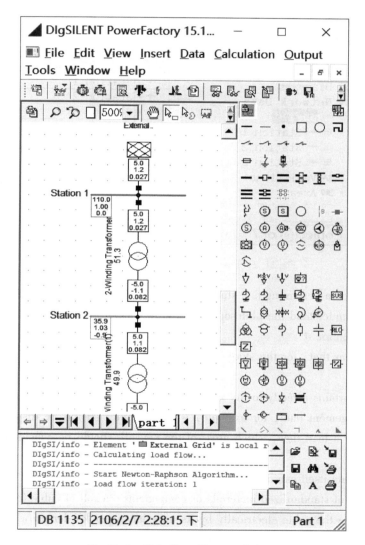

Fig. 11.1 Main PowerFactory windows

As users progress and become more adept with the program, data will be manipulated by using a data viewer called the Data Manager. The two means of accessing the data are thus via the graphics page and via the Data Manager.

Data is accessed from the graphics page by double-clicking on an object. An input dialogue pops up and the user may then edit the data for that object.

All of the data that is entered for such objects is hierarchically structured in folders to allow the user to navigate through it. To view the data and its organization, a Data Manager is used. Fig. 11.2 shows the Data Manager Window. The Data Manager is similar in appearance and working to a Windows Explorer.

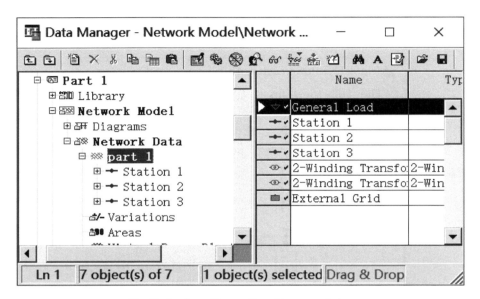

Fig. 11.2 PowerFactory Data Manager window

As mentioned, data pertaining to a study is organized into several folders. Before examining this structure we should understand the philosophy behind this arrangement.

5. Data Arrangement

Firstly, it is clear that, for the study of any system, there are two distinct sets of information that can be defined:

1) Data that pertains directly to the system under study, that is, electrical data.

2) Study management data, for example, which graphics should be displayed, what options have been chosen for a load flow, which areas of the network should be considered for calculation, etc.

The electrical data itself can also be further divided into logical sets. When we construct a power system we make use of standardized materials or components—a roll of cable for example. In simple terms we can describe the cable electrically by its impedance per km length whilst it is still on the cable drum; in other words, generic information about this cable, is called Type data.

When we cut a length of the cable for installation the type data is retained in a modified way, as follows: 600 m of cable that has a Type impedance of $Y\Omega/km$ will now have an impedance of $0.6Y\Omega$.

We can thus see that the length of the cable, 0.6 km, can be seen as a separate set of information. This set will contain all of that information particular to the specific installation or application of the piece of cable we are considering. Information such as the derating factor of the installed cable, its local name, the nodes that it is connected to at either end; in other words, all information that is non-generic, will fall into this information set. In PowerFactory we call this Element Data.

Within the Element Data, there is information related to the operational point of a devise but not to the devise itself, i. e. the tap position on a transformer or the active power dispatch of a generator. These kind of data, which is subject to frequent changes during a study and may be used to simulate different operation scenarios of the same network, is further grouped inside the Element Data set in a subset called Operational Data.

This means that there are now four distinct sets that we need to arrange the data into. In Database terms this means four folders, which, in PowerFactory, we call:

1) Network Data folder: Holds all the element data.

2) Operation Scenario folder: Holds the operational data defining a certain operational point.

3) Equipment Type folder: Holds all the type data.

4) Study Case folder: Holds all the study management data.

For an optimal advantage of the flexibility offered by this data arrangement approach, the aforementioned folders should be hierarchically organized within a higher directory. In PowerFactory this higher directory is called Project. Besides the described data sets, a project stores all the additional database "objects" required to model, simulate, analyze and visualize a particular power system.

The PowerFactory database supports multiple users and each user can manage several projects. User Account folders with access privileges only for their owners (and other users with shared rights) must then be used. User accounts are of course in a higher level than projects.

Fig. 11.3 shows a snapshot from a database as seen by the user in a data manager window. The folders listed contain the following type of data:

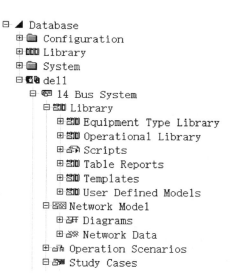

Fig. 11.3 Structure of a PowerFactory project in the Data Manager

1) User Folder: Three user accounts (Demo, Manual and Peter) containing different projects are shown.

2) Project: In this case named Simple Power System. This folder is the frame for all project subfolders.

3) Equipment Types: Holds all the Type data.

4) Network Model: Holds the Network Data folder containing the element data. The Network Model folder also contains the graphical objects folders (Diagrams), storing network diagrams and the network modifications folders (Variations) storing expansions or topological changes to be

applied in the original networks.

5) Operation Scenarios: Holds sets of operational data.

6) Study Cases: Contains the study Management Data—the tools and tool settings that are used to perform the calculations and the visualization of the results.

6. User Interface

The PowerFactory windows are the users interface to the program and the means to enter or manipulate data and/or graphics. DIgSILENT PowerFactory uses several kinds of windows some of which have been shown previously. To follow the explanation, please see Fig. 11.4.

1) The main PowerFactory window is described in the title bar—DIgSILENT PowerFactory 15.1.

2) The main menu bar contains the drop down menu selections.

3) The Graphical Editor displays single line diagrams, block diagrams and/or simulation plots of the current project. Studied networks and simulation models can be directly modified from the graphical editor by placing and connecting elements.

4) The Data Manager is the direct interface with the database. It is similar in appearance and working to a Windows Explorer. The left pane displays a symbolic tree representation of the complete database. The right pane is a data browser that shows the content of the currently selected folder.

5) When an object is right clicked (in the graphical editor or in the data manager) a context sensitive menu with several possible actions appears.

6) When an object is double clicked its edit dialogue pops up. The edit dialogue is the interface between an object and the user. The parameters defining the object are accessed trough this edit dialogue. Normally an edit dialogue is composed of several pages (also called tabs). Each tab groups parameters that are relevant to a certain function. In the example of Fig. 11.4 the Load Flow tab of a generator is shown, therefore only the generator parameters relevant to a load flow calculation are available.

7) At the bottom of the PowerFactory window, an output window with its own toolbar is shown.

The Data Manager sub-window (this window is created by pressing the icon, which is the first icon on the left of the main toolbar) is always floating and more than one can be active at the same time. The database manager itself has several appearances: It may only show the database tree for selecting a database folder, or it may be the full version with the database tree, the data browser, and all editing capabilities.

One of the major tasks for the data manager is to provide access to the power system components. The power system components shown in the data manager can be gang-edited (or group-edited) within the data manager itself, where the data is presented in a tabular format, for all the selected objects. Alternatively each object may also be individually edited by double clicking on an object (or right click→Edit).

The output window, at the bottom of the screen, is always there; it cannot be closed although it can be minimized. The output window can be docked, that is, fixed to a location on the bottom of the main window. The docked state is the default, as shown in the Fig. 11.4.

When clicking the right mouse button, when the cursor is in the output window area, the

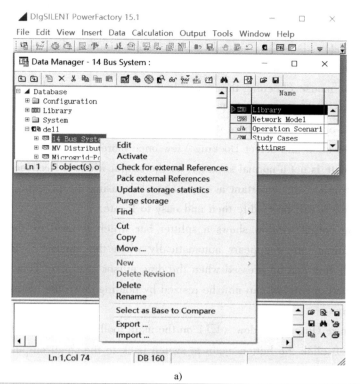

a)

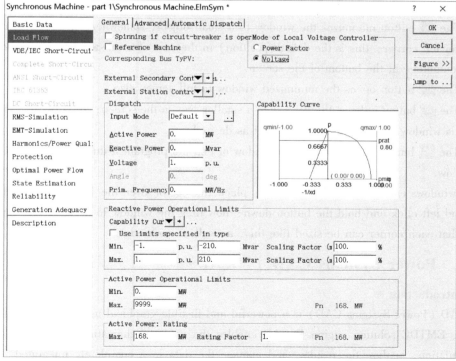

b)

Fig. 11.4 PowerFactory initial appearance

context sensitive menu of the output window appears. Then the output window can be undocked by deselecting the Docking View (by clicking the mouse onto Docking View to untick it). The undocked output window is still confined to the main window, but now as a free floating window. This sometimes occurs accidentally when the user left clicks the tool bar for the output window and drags the mouse (keeping the mouse button down) to somewhere outside of the output window boundaries. To rectify this simply left click in the title bar of the undocked window and drag it down to the bottom of the screen where it will dock once more (if you have right clicked unticked Docking View to right click and select Docking View once more.

The undocked state is not a normal situation for the output window. Because the output messages that appear in this window are important at any stage while using the program, the docked state is the best place because it will be visible then and easy to locate.

The edge of the output window shows a splitter bar which is used to change the size of the output window. The drag cursor appears automatically when the cursor is placed on the splitter bar. The left mouse button can be pressed when the drag cursor is visible. This will turn the splitter bar to grey and the output window can now be resized by holding down the mouse button and moving the mouse up or down.

The icon Maximize Output Window ( ) on the main toolbar will enlarge the output window to almost full-screen. Left click the button again to switch back to the small output window.

On the right of the title bar of any window or sub-window there are three buttons that are used to Minimize, Maximize/Restore Down or Close the window.

1) The button minimizes the window to a small object, somewhere on the screen (usually in the lower left corner—this is the default position) in the case of a sub-window, or to the task bar for your computer—at the bottom of the screen.

2) The button opens the minimized window to full screen size.

3) The button reduces the window to a smaller size on the screen; initially there is a default size for this window but the user may resize it as desired.

4) The button will close the sub-window or end the program if this button is clicked on the main window.

Sub-windows can be resized as follows: place the cursor over the lower right corner of the window and left click and hold the button down—now drag the window to the size you require. You will find that each corner can be sized like this, as well as each edge.

11.2.3 Power Systems CAD (PSCAD)

1. Introduction

PSCAD (Power Systems CAD) is a powerful and flexible graphical user interface to the world-renowned, EMTDC solution engine. PSCAD enables the user to schematically construct a circuit, run a simulation, analyze the results, and manage the data in a completely integrated, graphical environment. Online plotting functions, controls and meters are also included, so that the user can alter system parameters during a simulation run, and view the results directly.

PSCAD comes complete with a library of pre-programmed and tested models, ranging from simple passive elements and control functions, to more complex models, such as electric machines, FACTS devices, transmission lines and cables. If a particular model does not exist, PSCAD provides the flexibility of building custom models, either by assembling them graphically using existing models, or by utilizing an intuitively designed Design Editor.

The following are some common models found in systems studied using PSCAD:

1) Resistors, inductors, capacitors;
2) Mutually coupled windings, such as transformers;
3) Frequency dependent transmission lines and cables (including the most accurate time domain line model in the world);
4) Current and voltage sources;
5) Switches and breakers;
6) Protection and relaying;
7) Diodes, thyristors and GTOs;
8) Analog and digital control functions;
9) AC and DC machines, exciters, governors, stabilizers and inertial models;
10) Meters and measuring functions;
11) Generic DC and AC controls;
12) HVDC, SVC, and other FACTS controllers;
13) Wind source, turbines and governors.

PSCAD, and its simulation engine EMTDC, have enjoyed close to 30 years of development, inspired by ideas and suggestions by its ever strengthening, worldwide user base. This development philosophy has helped to establish PSCAD as one of the most powerful and intuitive CAD software packages available.

PSCAD can represent electric circuits in detail not available with conventional network simulation software. For example, transformer saturation can be represented accurately on PSCAD and only superficially, if at all, on Phasor based simulators such as power system stability programs.

It is obvious from this example that the instantaneous solution algorithm of PSCAD and the precision achievable with it opens up great opportunities for investigation and study. PSCAD is used by engineers, researchers and students from utilities, manufacturers, consultants, research and academic institutes. It is used in planning, design, developing new concepts, testing ideas, understanding what happened when equipment failed, commissioning, preparation of specification and tender documents, teaching and research.

PSCAD/EMTDC is under continual development by a team of engineers and computer scientists under the leadership of Garth Irwin. The development is guided by the needs of the many users around the world.

2. Typical PSCAD Studies

The PSCAD users' spectrum includes engineers and scientists from utilities, manufacturers, consultants, research and academic institutions. It is used in planning, operation, design, commissioning, preparing of tender specifications, teaching and research. The following are examples of types of studies routinely conducted using PSCAD:

1) Contingency studies of AC networks consisting of rotating machines, exciters, governors, turbines, transformers, transmission lines, cables, and loads;

2) Relay coordination;

3) Transformer saturation effects;

4) Insulation coordination of transformers, breakers and arrestors;

5) Impulse testing of transformers;

6) Sub-synchronous resonance (SSR) studies of networks with machines, transmission lines and HVDC systems;

7) Evaluation of filter design and harmonic analysis;

8) Control system design and coordination of FACTS and HVDC; including STATCOM, VSC, and cycloconverters;

9) Optimal design of controller parameters;

10) Investigation of new circuit and control concepts;

11) Lightning strikes, faults or breaker operations;

12) Steep front and fast front studies;

13) Investigate the pulsing effects of diesel engines and wind turbines on electric networks;

14) Incorporate the capabilities of MATLAB/Simulink directly into PSCAD/EMTDC;

15) Effects of DC currents and geomagnetically induced currents on power systems, inrush effects and ferroresonance;

16) Distribution system design with custom power controllers and distributed generation;

17) Power quality analysis and improvement;

18) Design, control coordination and system integration of wind farms, diesel systems, and energy storage;

19) Variable speed drives, their design and control.

11.2.4 Power System Analysis Toolbox (PSAT)

1. Introduction

Software packages for power system analysis can be basically divided into two classes of tools: commercial softwares and educational/research-aimed softwares. Commercial software packages available on the market (e.g. PSS/E, EuroStag, Simpow, and CYME) follows an all-in-one philosophy and are typically well-tested and computationally efficient. Despite their completeness, these softwares can result cumbersome for educational and research purposes. Even more important, commercial softwares are closed, i.e., do not allow changing the source code or adding new algorithms.

For research purposes, the flexibility and the ability of easy prototyping are often more crucial aspects than computational efficiency. On the other hand, there is a variety of open source research tools, which are typically aimed to a specific aspect of power system analysis. An example is

UWPflow, which provides an extremely robust algorithm for continuation power flow analysis. However, extending and/or modifying this kind of scientific tools also requires keen programming skills, in addition to a good knowledge of a low level language (C in the case of UWPflow) and of the structure of the program.

In the last decade, several high-level scientific languages, such as MATLAB, Mathematica and Modelica, have become more and more popular for both research and educational purposes. Any of these languages can lead to good results in the field of power system analysis. However MATLAB proved to be the best user choice. Key features of MATLAB are the matrix-oriented programming, excellent plotting capabilities and a graphical environment (Simulink) which highly simplifies control scheme design. For these reasons, several MATLAB-based commercial, research and educational power system tools have been proposed, such as power system toolbox (PST), MatPower, Toolbox (VST), MatEMTP, SimPowerSystems (SPS), power analysis toolbox (PAT), and the educational simulation tool (EST). Among these, only MatPower and VST are open source and freely downloadable.

This section describes a MATLAB-based power system analysis tool (PSAT) which is freely distributed on line. PSAT includes power flow, continuation power flow, optimal power flow, small-signal stability analysis, and time-domain simulation. The toolbox is also provided with a complete graphical interface and a Simulink-based one-line network editor. Tab. 11.1 depicts a rough comparison of the currently available MATLAB-based tools for power system analysis and PSAT.

Tab. 11.1 MATLAB-based packages for power system analysis

Package	PF	CPF	OPF	SSA	TD	EMT	GUI	GNE
EST				√	√			√
MatEMTP					√	√	√	√
MatPower	√		√					
PAT	√			√	√			√
PSAT	√	√	√	√	√		√	√
PST	√	√		√	√			
SPS	√			√	√	√	√	√
VST	√	√		√	√		√	

The features illustrated in the Tab. 11.1 are the power flow (PF), the continuation power flow and/or voltage stability analysis (CPF-VS), the optimal power flow (OPF), the small signal stability analysis (SSA), and the time-domain simulation (TD) along with aesthetic features such as the graphical user interface (GUI) and the graphical network editor (GNE).

An important but often missed issue is that the MATLAB environment is a commercial and closed product, thus MATLAB kernel and libraries cannot be modified nor freely distributed. To allow exchanging ideas and effectively improving scientific research, both the toolbox and the platform on which the toolbox runs should be free. At this aim, PSAT can run on GNU/Octave, which is a free MATLAB clone.

2. PSAT Features

(1) Outlines

PSAT has been thought to be portable and open source. At this aim, PSAT has been developed using MATLAB, which runs on the commonest operating systems, such as UNIX, Linux, Windows and Mac OS X. Nevertheless, PSAT would not be completely open source if it run only on MATLAB, which is a proprietary software. At this aim PSAT can run also on the latest GNU/Octave releases, which is basically a free MATLAB clone. In the knowledge, PSAT is actually the first free software project in the field of power system analysis. PSAT is also the first power system software which runs on GNU/Octave platforms.

The synoptic scheme of PSAT is depicted in Fig. 11.5.

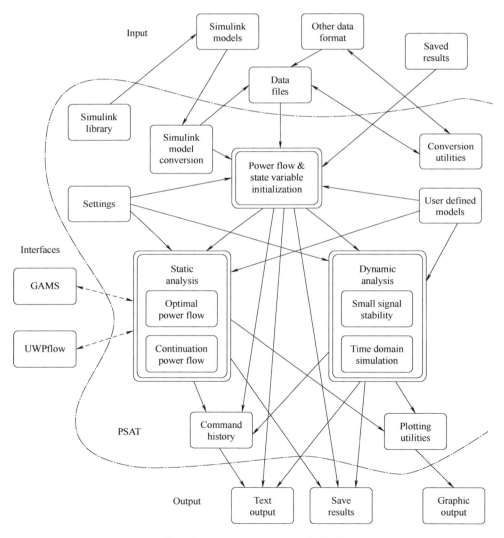

Fig. 11.5 Synoptic scheme of PSAT

Observe that PSAT kernel is the power flow algorithm, which also takes care of the state variable initialization. Once the power flow has been solved, the user can perform further static and/or dynamic analyses. These are:

1) Continuation power flow (CPF);
2) Optimal power flow (OPF);
3) Small signal stability analysis;
4) Time domain simulations.

PSAT deeply exploits MATLAB vectorized computations and sparse matrix functions in order to optimize performances. Furthermore PSAT is provided with the most complete set of algorithms for static and dynamic analyses among currently available MATLAB-based power system softwares (see Tab. 11.1).

PSAT also contains interfaces to UWPflow and GAMS which highly extend PSAT ability to solve CPF and OPF problems, respectively. These interfaces are not discussed here, as they are beyond the main purpose of this section.

In order to perform accurate and complete power system analyses, PSAT supports a variety of static and dynamic models, as follows:

1) Power flow data: bus bars, transmission lines and transformers, slack buses, PV generators, constant power loads, and shunt admittances.

2) Market data: power supply bids and limits, generator power reserves, and power demand bids and limits.

3) Switches: transmission line faults and breakers.

4) Measurements: bus frequency measurements.

5) Loads: voltage dependent loads, frequency dependent loads, ZIP (polynomial) loads, thermostatically controlled loads, and exponential recovery loads.

6) Machines: synchronous machines (dynamic order from 2 to 8) and induction motors (dynamic order from 1 to 5).

7) Controls: turbine governors, AVRs, PSSs, over-excitation limiters, and secondary voltage regulation.

8) Regulating transformers: under load tap changers and phase shifting transformers.

9) FACTS: SVCs, TCSCs, SSSCs, UPFCs.

10) Wind turbines: wind models, constant speed wind turbine with squirrel cage induction motor, variable speed wind turbine with doubly fed induction generator, and variable speed wind turbine with direct drive synchronous generator.

11) Other models: synchronous machine dynamic shaft, subsynchronous resonance model, solid oxide fuel cell, and subtransmission area equivalents.

Besides mathematical algorithms and models, PSAT includes a variety of additional tools, as follows:

1) User-friendly graphical user interfaces;
2) Simulink library for one-line network diagrams;

3) Data file conversion to and from other formats;
4) User defined model editor and installer;
5) Command line usage.

The following subsections will briefly describe these tools. Observe that, due to GNU/Octave limitations, not all algorithms/tools are available on this platform (see Tab. 11.2).

Tab. 11.2　Functions available on MATLAB and GNU/Octave platforms

Function	MATLAB	GNU/Octave
Continuation power flow	yes	yes
Optimal power flow	yes	yes
Small signal stability analysis	yes	yes
Time domain simulation	yes	yes
GUIs and Simulink library	yes	no
Data format conversion	yes	yes
User defined models	yes	no
Command line usage	yes	yes

(2) Getting Started and Main Graphical User Interface

PSAT is launched by typing at the MATLAB prompt:

```
>> psat
```

which will create all structures required by the toolbox and open the main GUI (see Fig. 11.6). All procedures implemented in PSAT can be launched from this window by means of menus, buttons and/or short cuts.

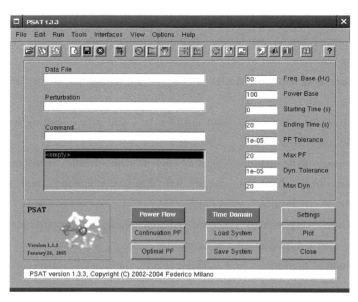

Fig. 11.6　Main graphical user interface of PSAT

The main settings, such as the system base or the maximum number of iteration of Newton-Raphson methods, are shown in the main window. Other system parameters and specific algorithm settings have dedicated GUIs. Observe that PSAT does not rely on GUIs and makes use of global variables to store both setting parameters and data. This approach allows using PSAT from the command line as needed in many applications.

3. **Simulink Library**

PSAT allows drawing electrical schemes by means of pictorial blocks. Fig. 11.7 depicts the complete PSAT-Simulink library.

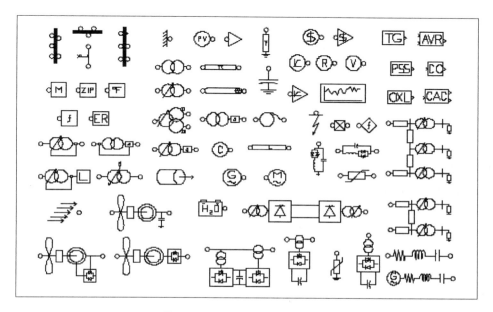

Fig. 11.7　PSAT Simulink library

The PSAT computational engine is purely MATLAB-based and the Simulink environment is used only as graphical tool. As a matter of fact, Simulink models are read by PSAT to exploit network topology and extract component data. A byproduct of this approach is that PSAT can run on GNU/Octave, which is currently not providing a Simulink clone.

Observe that some Simulink-based tools, such as PAT and EST, use Simulink to simplify the design of new control schemes. This is not possible in PSAT. However, PAT and EST do not allow representing the network topology, thus resulting in a lower readability of the whole system.

4. **Data Conversion and User Defined Models**

To ensure portability and promote contributions, PSAT is provided with a variety of tools, such as a set of data format conversion (DFC) functions and the capability of defining user defined models (UDMs).

The set of DFC functions allows converting data files to and from formats commonly in use in power system analysis. These include: IEEE, EPRI, PTI, PSAP, PSS/E, CYME, MatPower and PST formats. On MATLAB platforms, an easy-to-use GUI (see in Fig. 11.8) handles the DFC.

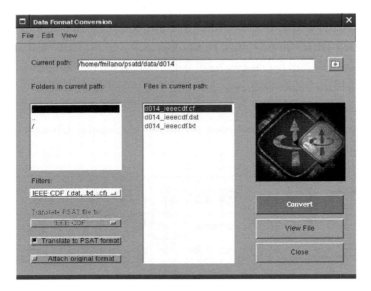

Fig. 11.8 GUI for data format conversion

The UDM tools allow extending the capabilities of PSAT and help end-users to quickly set up their own models. UDMs can be created by means of the GUI depicted in Fig. 11.9. Once the user has introduced the variables and defined the DAE of the new model in the UDM GUI, PSAT automatically compiles equations, computes symbolic expression of Jacobians matrices (by means of the Symbolic Toolbox) and writes a MATLAB function of the new component. Then the user can save the model definition and/or install the model in PSAT. If the component is not needed any longer it can be uninstalled using the UDM installer as well.

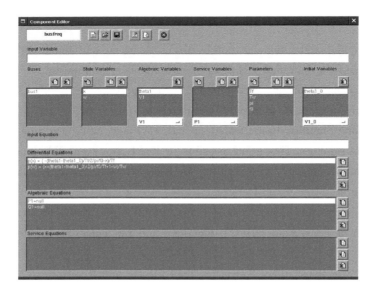

Fig. 11.9 GUI for user defined models

5. Command Line Usage

GUIs are useful for education purposes but can in some cases limit the development or the usage of a software. For this reason PSAT is provided with a command line version. This feature allows using PSAT in the following conditions:

1) If it is not possible or very slow to visualize the graphical environment (e. g. MATLAB is running on a remote server).

2) If one wants to write scripting of computations or include calls to PSAT functions within user defined programs.

3) If PSAT runs on the GNU/Octave platform, which currently neither provides GUI tools nor a Simulink-like environment.

11.3 Real-Time Digital Simulator (RTDS)

11.3.1 Introduction

The real-time digital simulator (RTDS) is a special purpose computer designed to study electromagnetic transient phenomena in real-time. The RTDS is comprised of both specially designed hardware and software. RTDS hardware is digital signal processor (DSP) and reduced instruction set computer (RISC) based, and utilizes advanced parallel processing techniques in order to achieve the computation speeds required to maintain continuous real-time operation.

RTDS software includes accurate power system component models required to represent many of the complex elements which make up physical power systems. The overall network solution technique employed in the RTDS is based on nodal analysis.

The underlying solution algorithms are those introduced in the now classic paper *Digital Computer Solution of Electromagnetic Transients in Single and Multiphase Networks* by H. W. Dommel. Dommel's solution algorithm is used in virtually all digital simulation programs designed for the study of electromagnetic transients.

RTDS software also includes a powerful and user friendly graphical user interface (GUI), referred to as RSCAD, through which the user is able to construct, run and analyze simulation cases.

11.3.2 RTDS Hardware

Unlike analogue simulators, which output continuous signals with respect to time, digital simulators compute the state of the power system model only at discrete instants in time. The time between these discrete instants is referred to as the simulation time-step (Δt). Many hundreds of thousands of calculations must be performed during each time-step in order to compute the state of the system at that instant. The temporary transients class of studies for which the RTDS is most often used requires Δt to be in the order of 50 to 60 μs (frequency response accurate to approximately

3000 Hz). By definition, in order to operate in real-time a 50 μs time-step would require that all computations for the system solution be complete in less than 50 μs of actual time.

In order to realize and maintain the required computation rates for real-time operation, many high speed processors operating in parallel are utilized by the RTDS. Two types of processor cards may be installed in each RTDS rack.

The triple processor card (3PC) contains three analogue devices ADSP 21062 digital signal processors. The ADSP 21062 DSP clock speed is 40 MHz.

The risc processor card (RPC) contains two PowerPC 750CXe risc processors operating at a clock speed of 600 MHz. The specific PowerPC supplied with your RTDS simulator may not be a 750CXe, as more powerful versions are used as they become available.

The RTDS simulator can be configured as 3PC only or as a combination of 3PC and RPC.

A rack of RTDS hardware is defined as one housing consisting of up to 20 printed circuit boards. Typical configurations include the following number of processor cards:

12 × 3PC,

8 × 3PC, 1 × RPC

In addition to processing cards, an RTDS rack always contains a workstation interface card and in the case of multi-rack systems, an inter-rack communications card.

1. The RISC Processor Card (RPC)

As mentioned, the RPC card contains 2 PowerPC 750CXe processors, each mounted on a daughter card subassembly. The cards communicate through a local high speed ring bus that also includes the back plane. The RPC has no indpendent I/O facilities in the current version, and is used primarily for the rack network solution.

2. The Triple Processor Card (3PC)

Each triple processor card contains three independent ADSP 21062 processors and their associated memory, backplane interface and input/output ports. Each 3PC contains the following I/O ports:

24 × analogue output channels (12 bit +/−10 V range);

2 × digital input port (16 bit each);

2 × digital output port (16 bit each).

The I/O capabilities of the three processors (A, B and C) are summarized follows. Each of the three processors (A, B and C) has access to eight analogue output channels. In addition, processors A and B each have access to one digital input port and one digital output port. No digital ports are associated with processor C. Processor C is used for optionally available analogue I/O channels. Application of the I/O ports varies depending upon the type of power system component model which has been assigned to run on the 3PC processors.

In order to import analogue signals to a 3PC card, optional auxiliary hardware is required. The optional OADC is used together with a 3PC card to achieve analogue input. Six independent input channels are available on each OADC (+/−10 V peak, optically isolated, differential inputs). When included, OADC boards are rail mounted in the rear of the RTDS cubicle and connected to a

3PC card using fiber optic cable. Processor C of the 3PC card can be optionally fitted with the fiber optic signal receiver when analogue input capability is required. For each OADC board installed in the RTDS, one 3PC is equipped with the fiber optic connection hardware.

3. Workstation Interface Card (WIF)

One WIF card is installed per RTDS rack. Each WIF performs four main functions:

1) Rack diagnostics;
2) RTDS-to-computerworkstation communications;
3) Multi-rack case synchronization;
4) Backplane communications.

Rack diagnostics are running whenever the rack power is turned on or when the WIF front panel RST button is pushed. Results from the diagnostics can be accessed using RSCAD/RunTime.

Communication between the RTDS and the host computer workstation is done using a 10/100 Base-T Ethernet link. The WIF may be directly connected to a host workstation using a swap type cable, or the WIF may be connected to the local area network using an Ethernet hub.

The WIF is responsible for maintaining synchronization between individual racks in a multi-rack simulation case. Simulators which include 3 or more racks require a global bus hub (GBH). The GBH is installed in the rear of one of the RTDS cubicles and is used to facilitate direct communication of certain signals between RTDS racks during a simulation. RTDS simulators which consist of only one rack do not require a GBH. The WIF cards in a two rack RTDS are directly connected using fiber optic cable and do not need a GBH. The RTDS hardware manual contains a complete explanation of the WIF and the GBH.

Within a single rack many signals are exchanged between 3PC, WIF and IRC cards along a common communication backplane. Each card is directly connected to the communication backplane. The WIF card is responsible for coordinating backplane communication.

4. Inter-Rack Communications Card (IRC)

One inter-rack communication card is installed in each rack of a multi-rack RTDS simulator. The IRC is used to communicate data between interconnected racks. High speed parallel to serial and serial to parallel data converters allow connections to be made between racks. Each IRC includes six bidirectional data communication paths. An RTDS simulator consisting of seven racks can thus have direct communication between all racks.

A pair of front panel LEDs exists for each of the six communication channels on the IRC. The green LED, when on, indicates that the associated channel is active for the simulation case. The LED will only come on when a simulation case is running and if the RTDS software has determined that direct communication between the two racks connected by the channel is needed. The red LED, when on, indicates that an invalid data packet was received. The serial communication protocol between the sending and receiving includes extra bits for error detection (not correction). A single transmission error will cause the red LED to stay on until the simulation case is stopped.

11.3.3 RTDS Software

Software for the RTDS is organized into a hierarchy containing three separate levels: high-level graphical user interface, mid-level compiler and communications, and the low-level WIF multi-tasking operating system. The RTDS user is exposed only to the high-level software with the lower levels being automatically accessed through higher level software.

1. RSCAD Graphical User Interface

The high-level RTDS software comprises the RSCAD family of tools. RSCAD is a software package developed to provide a fully graphical interface to the RTDS. Prior to the development of RSCAD, another software suite—PSCAD served as the graphical user interface to the RTDS hardware.

RSCAD/FileManager (Fileman) represents the entry point to the RSCAD interface software. Fileman is used for project and case management and facilitates information exchange between RTDS users. All other RSCAD programs are launched from the Fileman module.

RSCAD/Draft is used for circuit assembly and parameter entry. The Draft screen is divided into two sections: the library section and the circuit assembly section. Individual component icons are selected from the library and placed in the circuit assembly section. Interconnection of individual component icons and parameter entry follows through a series of menus.

RSCAD/T-Line and RSCAD/Cable are used to define the properties of overhead transmission lines and underground cables respectively. Data is generally entered in terms of physical geometry and configuration. Line and Cable constants and equations are solved, resulting in ready to use data for the RSCAD/Draft program. Draft cross-references line and cable output files by name.

RSCAD/RunTime is used to control the simulation case(s) being performed on the RTDS hardware. Simulation control, including start/stop commands, sequence initiation, set point adjustment, fault application, breaker operation, etc. are performed through the RunTime Operator's Console. Additionally, on-line metering and data acquisition/disturbance recording functions are available in RunTime.

RSCAD/MultiPlot is used for post processing and analysis of results captured and stored during a simulation study. Report ready plots can be generated by MultiPlot.

2. RTDS Compiler/Linker & Operating System

RTDS mid-level software is divided into two separate areas: the operating system and the compiler. Although neither of these elements of the RTDS software are directly accessed by the user, some mention of their role in the overall software structure must be given in order to provide a better understanding of RTDS operating principles.

The RTDS operating system performs many functions. Part of the O/S runs on the host computer workstation while part runs on the workstation interface cards. The major function of the WIF based portion of the O/S is to handle I/O requests which are usually initiated by the user. Diagnostic tests performed by the system administrator are also handled through the operating system level of software residing on the WIF. Finally, cross-rack communication errors, if they occur, will be detected by the WIF based operating system software which will in turn cause the simulation to be stopped and the appropriate LED indicators to be illuminated.

Generation of executable code required for each new simulation case is done through a specially developed set of software programs collectively termed the RTDS compiler.

The compiler takes as input the power system data entered by the user through RSCAD/Draft along with a hardware configuration file (RSCAD\HDWR\config_file) which defines the hardware making up the user's RTDS installation. As output, the compiler produces all of the parallel processing code required by the digital signal processors, as well as memory allocation and data communication transfer schedules.

In order to generate the executable code, the compiler accesses the lowest level of RTDS software—the component library. The library contains code modules for all available power and control system component models. The code modules generally consist of low-level machine language code for the individual component models and also for the overall system solution. Based on the user defined circuit, processor allocation and required library access will be performed by the RTDS compiler in a manner transparent to the user. Processor assignment can be either automatic (i. e. decided by the compiler) or can be manually specified by the user during the RSCAD/Draft session. The final product of the compiling process is a file (or set of files) containing DSP code which is transferred to the RTDS over the Ethernet using the RSCAD/RunTime start command.

The compiler also produces a . MAP file. The . MAP file is a user readable file which provides information on processor allocation (i. e. cross reference listing for component/processor match-up), input/output channel allocation, analogue output channel scaling and system initial conditions. The . MAP file is particularly useful and important when physical connections are to be made between the RTDS and external equipment.

3. **RTDS Simulator Power System Component Library**

As explained in the RSCAD/Draft manual, icons representing all available RTDS simulator models are stored in one or more libraries. Several libraries have been included in the RSCAD software installation. These libraries are stored at the "Master" level within RSCAD/Draft.

In general the libraries supplied by RTDS technologies are separated into the following categories:

1) 3PC Power System Components;
2) 3PC Control System Components;
3) 3PC IEEE Generator Control Components;
4) 3PC Complex Control Components;
5) Load Flow Components/Single Line Diagrams (LF_SLD).

The User can select any one of these libraries during a RSCAD/Draft session. The 3PC based power system component models can, for example be accessed from RSCAD/Draft by using: File→Open→Library, selecting Master, and then choosing the 3PC_Power_System tab near the top of the Library window. In addition to the supplied libraries, the User is able to create and customize additional libraries and store them in the "User" level of RSCAD/Draft by clicking on the New Tab button on the library button bar. This creates a new library, which appears as a blank tab in the library window. The blank library tab may then be saved by right-clicking it, and choosing Save Tab As from the popup menu that appears. The User will then be prompted to provide a name for the library.

Components may be added to a library by right-clicking on the library canvas and choosing Add Component. The component to be added can then be located by first choosing the library type that contains it (User or Master), and then navigating to that folder. Once located, the component may be added by double clicking on it, or clicking Open with the component's file selected. Alternatively, components may be copied (or moved) into a library from another library by right clicking the component and choosing copy (or move) from the popup window. The component may then be placed into the desired library by clicking on the desired library canvas. Note that the copy operation results in a duplicate of the component being placed, while the move operation results in the relocation of the component (it is removed from its original location).

VOCABULARY

1. graphical   *adj.* 绘画的；生动的；图解的
2. auxiliary   *n.* 助动词；辅助者，辅助物；附属机构   *adj.* 辅助的；副的；附加的
3. manipulation   *n.* 操作；操纵；处理；篡改
4. intimate   *adj.* 亲密的；私人的   *n.* 知己   *vt.* 暗示；通知；宣布
5. licensed   *adj.* 得到许可的（等于 licenced）   *v.* 许可；批准（license 的过去分词）
6. interactive   *adj.* 交互式的；相互作用的
7. implemented   *v.* 实施（implement 的过去分词形式）；执行   *adj.* 应用的
8. memorization   *n.* 记住；暗记
9. tedious   *adj.* 沉闷的；冗长乏味的
10. scenarios   *n.* 情节；脚本；情景介绍（scenario 的复数）
11. aforementioned   *adj.* 上述的；前面提及的
12. hierarchically   *adv.* 分层次，分等级地
13. snapshot   *n.* 快照，快相   *vt.* 给…拍快照   *vi.* 拍快照
14. superficially   *adv.* 表面地；浅薄地
15. cycloconverter   *n.* 回旋转换器，交-交变频器
16. geomagnetically   *adv.* geomagnetic 的变形
17. geomagnetic   *adj.* 地磁的；地磁场的
18. ferroresonance   *n.* 铁磁谐振，铁谐振
19. diesel   *n.* 柴油机；柴油   *adj.* 内燃机传动的；供内燃机用的
20. sparse   *adj.* 稀疏的；稀少的
21. polynomial   *n.* [数]多项式；由两个字以上组成的学名   *adj.* 多项式的；多词学名（指由两个以上的词构成的学名）
22. thermostatically   *adv.* 恒温地；自动调节温度地
23. subtransmission   *n.* 二次输电；辅助变速箱；中等电压输电
24. Newton-Raphson   牛顿-拉夫逊
25. hierarchy   *n.* 层级；等级制度

NOTES

1. PSS/E is composed of a comprehensive set of programs for studies of power system

transmission network and generation performance in both steady-state and dynamic conditions.

电力系统工程仿真（PSS/E）由一套综合性的程序包组成，用于研究电力系统输电网和发电机在稳态和动态条件下的性能。

2. Software packages for power system analysis can be basically divided into two classes of tools: commercial software and educational/research-aimed software.

电力系统分析软件包基本分为两类工具：商业软件和教育/研究用软件。

Reference

[1] VITTA L V, MCCALLEY J, AJJARAPU V, et al. Impact of increased DFIG wind penetration on power systems and markets [R]. Kalgary: Power Systems Engineering Research Center Texas A&M University, 2005.

[2] XU B, ABUR A. Optimal placement of phasor measurement units for state estimation [R]. Kalgary: Power Systems Engineering Research Center, Texas A&M University, 2005.

[3] PRABHA K. Power system stability and control [M]. New York: McGraw-Hill Professional, 1994.

[4] GLOVER J D, MULUKUTIA S S, THOMAS J O. Power system analysis and design [M]. Stamford: Cengage Learning, 2016.

[5] Power Systems Engineering Research Center. Challenges in integrating renewable technologies into an electric power system: White Paper [R]. Phoenix: Arizona State University, 2010.

[6] Power Systems Engineering Research Center. U. S. energy infrastructure investment: large-scale integrated smart grid solutions with high penetration of renewable resources, dispersed generation, and customer participation: White Paper [R]. Phoenix: Arizona State University, 2009.

[7] KEZUNOVIC M, GUAN Y F, GUO C Y, et. al. The 21st century substation design [R]. Phoenix: Power Systems Arizona State University, 2010.

[8] OVERBYE T. Using PMU data to increase situational awareness [R]. Phoenix: Engineering Research Center, Arizona State University, 2010.

[9] HINGORANI N G. Flexible AC transmission [J]. IEEE Spectrum, 1993, 30 (4): 40-45.

[10] GYUGYI L, SCHAUDER C D, WILLIAMS S L, et al. Unified power flow controller: a new approach to power transmission control [J]. IEEE Transactions on Power Delivery, 1995, 10 (2): 1085-1097.

[11] HINGORANI N G. Flexible AC transmission [J]. IEEE Spectrum, 1993, 30 (4): 40-45.

[12] KEZUNOVIC M, GUO C, GUAN Y, et al. New concept and solution for monitoring and control system for the 21st century substation [C]. Hangzhou: International Conference on Power System Technology, 2010.

[13] KARLSSON D, BROSKI L, GANESAN S. Maximizing power system stability through wide area protection [C]. Texas: 57th Annual Conference for Protective Relay Engineers, 2004.

[14] 李久胜, 马洪飞, 陈宏钧, 等. 电气自动化专业英语: 修订版 [M]. 哈尔滨: 哈尔滨工业大学出版社, 2005.